Thirteen Pathways of Occult Herbalism

Thirteen Pathways of Occult Herbalism

and Other Homilies on Botanical Magic

Daniel A. Schulke

Illustrations by Benjamin A. Vierling

THREE HANDS PRESS

2017

First edition published by Three Hands Press, May 2017.

'Occult Herbalism: Ethos, Praxis, and Spirit-Congress' was originally
published in *The Cauldron* volume 127, February 2008. The version in the
present book has been revised and expanded. 'The Green Intercessor' was
originally published in *Abraxas* volume 1, 2009.

Jacket cover image *Witch's Garden* and rear cover images by
Benjamin A. Vierling. Jacket design by Bob Eames.
Interior book layout by Clint Marsh.

ISBN 978-1-945147-04-3 (softcover)

threehandspress.com

Contents

Preface

THE REPOSITORY of human-plant knowledge is prehistoric, and embraces arenas of understanding routinely rejected by science: the magical, mystical, spiritual, and mythical. These categories of knowledge are bodies of instruction and mystery that were hard won, over millennia, by technicians of magic: shamans, sorcerers, healers, and a thousand other specialists apprenticed to plants. Occult Herbalism, as it is presently used in this book, encompasses as a broad concept these zones of ritual botanical power.

Early in my youth, driven both by curiosity and by experiences inexplicable to science, philosophy, or religion, I began a personal investigation into this virid expanse of botanical power, a journey wending from wilderness to farm, from chapel to tumulus, from clinic to alchemical laboratory, even unto the veils of ecstatic trance and oneiric revelation. The fruits of this peregrination, in part, are contained within my book *Arcana Viridia: The Green Mysteries*, which began in the late 1980s as nothing more than a record of personal observations to better my own understanding.

Over the years, in my varied contacts with those interested in this subject, I was frequently asked 'how' one comes to such knowledge. Among contemporary students of esotericism there is considerable interest in how to approach plant powers from a respectful position, yet one that also allows the full possibilities of awakening to their mysteries. To these people I am thankful, for the dialogues thus incepted catalyzed the process leading to this book.

The purpose of *Thirteen Pathways* is therefore to examine routes by which we can learn of the occult nature of plants, and in doing so, incorporate their powers in our own mystical pursuits, and beyond. More than mere approaches, these pathways, when embodied, cohere a magical stance, a viewpoint which may readily be applied to any form of magic or spiritual approach, but also in everyday life.

In learning, acquisition of the simple facts of knowledge is simply not enough: our study must be immediated by personal insight, sharpened upon the grindstone of continual practice, fortified by willingness to be challenged and even proven wrong, but most importantly, the ability to strengthen one's personal path, and actuate knowledge by deed. In cases where the practitioner serves the community, such as offering a needed skill like midwifery or medicine, this actuation is something that cannot be faked.

The common English idiom of possessing a 'green thumb' or being 'green-fingered,' echoing old appellations of faerie, implies a secret bond of personal power with plants, particularly in horticulture. It also suggests that some people do not possess it, and are therefore excluded from plant power: the lack of this miraculous green digit is often given as reason for a strained or nonexistent connection with plants. This enduring perception of a relationship to plants as being a knack, a rare gift, or as the practice of an elite, resonates strongly with the concept of occultism and magic, for these

are extraordinary concepts, trafficking in rarity, conceal-
ment, and alienation. This book does not seek to render final
judgment on this matter, but it is a useful point of beginning,
in consideration of the rare and personal nature of occult
herbalism.

In common with *The Green Mysteries*, these writings find their
origin in learning, practice, and collaboration with others, but
also in time spent alone in the wilderness: places of isolation
and starkly alienating beauty, where for many years I have
deliberately sought solitude as an 'empty vessel.' Emergent
from a temporal trajectory passing through many circles and
fields, the present work, like that of the magical practitioner,
reminds us that knowledge is a process, a pilgrimage, that
does not cease once a thing is learned, but rather continues to
evolve and re-order the individual. The greater the degree of
practice and immersion in these matters, the more immediate
and profound our results will be. We seek, in short, the Stone
of the Alchemists: there is a tradition, some say heretical, that
its color is Green. The True Labor is neither easy, nor tolerant
of fools: Nature is a harsh mistress, and as the alchemists well
knew, one must be willing to pass through her disorienting
array of processes if one would glimpse beneath her veils.

Thirteen Pathways
of Occult Herbalism

MAGICAL OPERATIONS make their appearance in the earliest of human writing, and some of the most ancient inscriptions of mankind are formulae extolling the virtues of certain plants. A certain leaf is prescribed for the binding of a demon, or a specific root for making an animated statue; this knowledge is presented as authoritative and therefore worthy of preservation. We also accept that the use of plants in magical practices pre-dated writing systems, for this is increasing supported by archaeological evidence. What is more difficult to ascertain is the knowledge base that led to this ancient sorcery, the understanding of what gave plants their magical power, what spirits they embodied, what was required to work with them, and the correct manner to make use of their properties. This body of knowledge, which we might call a magical philosophy of trees and herbs, I refer to as Occult Herbalism. Though much of this elder knowledge is lost, most of these powerful plants are still with us, and despite the wreckage of civilizations, some of their traditions have been passed down through millennia, sometimes in the form of writing, and sometimes hand to hand, from master to apprentice.

It is tempting to conceive of Occult Herbalism based purely upon the more lurid and profane depictions of the occult arts, as they appear in popular culture: plants used for drugs, murder, and magic. We reject this characterization in the first instance because of its context: with the vast exposure and wide acceptance of a thing, or its reduction to entertainment, it ceases to be 'occult.' Occult, meaning 'hidden' is by its nature umbral, immaterial, private, encrypted, ineffable, mystical, and, importantly, concealed from the eyes of those who would abuse it.

One may also argue that the concerns of the plant world, by their nature, are 'occult' or 'esoteric' given their distance and state of estrangement from most human hearts. To many, the greensward is something to walk across, not to contemplate as a haven of lore and occult power. The cornfield similarly is an agrarian concern, abstracted from daily life and only conceptually related to sustenance, and the roses bought from the florist an ephemeral spot of color and fragrance serving to make a statement that words cannot. All of these things, however, have ancient associations and a related retinue of invisible powers, interweaving the spiritual and religious currents which feed the present. The pervasive state of apathy which often attends upon all matters vegetal has created shadows about them, and, in part, this has nourished their occult or hidden nature.

There is also, despite the legacies of the Age of Enlightenment, the persistence of magic and religion in the world, the traffic with divine power, and plants form an important part of this. In religion, herbs are powerfully crystallized in complex symbolism and theological narrative, as well as serving roles in the various rites of each canon. In magical practice, the study of plants has immediate applications in several established occult streams. Among the most prominent of these are traditions of spirit healing, or indigenous practices which outsiders call 'shamanism.' In the occult heritage of Europe, the strongest strands of esoteric botany occur in Alchemy and renaissance Natural Magic, which have several important schools specifically focusing on plant work, as well as witchcraft and herbal folk magic preserved at the local level. These systems are usually part of larger magical frameworks that include many other non-plant practices, such as angelic conjuration, planetary magic, kabbalah, and the corpus of Solomonic spirit-conjurations. As a discipline unto itself, Occult herbalism itself may also form the singular marrow of esoteric study and practice, focusing wholly on plants. In such cases the older exemplars of these teachings often do not define themselves as 'occult herbalism', rather one learns to become an 'herbalist,' or 'one who knows the secrets of plants' or 'herb-wise.'

In the course of study, the contemporary pupil of magic and occultism is often faced with plant references in the midst of a magical operation, even if it does not specifically concern plants; what is usually not apparent is the complex traditions which lie behind the herb and its acknowledged spiritual powers. In other cases, more cohesive bodies of occult plant doctrine present a bewildering array of teachings and lore, and the seeker naturally must consider how best to comprehend and implement this knowledge.

The model I propose represents an approach to learning, and it contains four essential features. The first of these are Pathways, of which I have for these purposes enumerated thirteen. There are also Gardens, for the purposes of this book accounted as thirteen, but their true number beyond count. The third feature of course is that of the seeker, the pilgrim in Elysium, and the fourth is the plants themselves. This formula represents a metaphysical model of a very physical process, a means by which the sublime power of plants can be approached in a meaningful and active way. The operation is dynamic, and ongoing, ever so much as the processes of Nature, which must be understood to unite its variables. In this, we resolutely identify with and thereby honor the axioms of Natural Magic.

The Pathways, as here exposited, are routes of approach to the mystery. Each presupposes a spiritual and philosophical stance, but also a momentum. In considering these pathways, it is important to note that each has a static emanation. Knowledge of the Pathway thus entails how it is expressed in motion, and also how it functions as a set of first principles. If we consider monuments in the landscape, the meaning of the Pathway becomes clearer: a mountain may be approached by many routes, affording different vistas; the mountain is singular but one's experience of it differs based on the road leading there.

If the Pathways indicate essential philosophical routes, the Gardens in turn are the zones of knowledge the Pathways lead to. Many of these overlap each other, and share arcana. The Gardens, thus, are concentrations of specified power; the Pathways are the routes leading there. Any Garden can thus be reached by one or more Pathways; likewise a single Pathway may perambulate multiple gardens. However, as all pilgrims know, a path may be trod in pursuit of a destination without arriving there: the path may turn, stray, or, by the nature of its demands, forcibly drive one to other by-ways, or into the thorny tangle of the wayside thicket.

The Pathways

Κάθαρσις • Katharsis
THE PATHWAY OF THE VIRGIN

All roads have their beginning, and that which penetrates the gardens of plant-mystery is no different, having a point of origination and emergence, if only in the flame of desire and aspiration. The recollection of early or 'first' experiences is universal: smelling the scents of certain flowers, the sudden and unexpected puncture of a thorn, tasting one's first cup of wine, and other altered states of consciousness brought on by plants. Aside from the innate characteristics of the thing encountered, the *tabula rasa* contributes power to this experience, the lack of individual epistemology more fully forming the experience of communion with the Other.

Thus is the Pathway of the Virgin—a road of our Art whose associated word *katharsis*, Greek for cleansing or purification, implies a state of the zeroth path, the point preceding all pathways. Admittedly the hallowed state of virginity is often scorned and ridiculed, regarded as naïve, inexperienced, and ignorant: these are but the bumptious out-gassings of the sexually atrophied. This fundamentally cowardly stance does not obtain within a magical framework, for it cannot admit that all things at one time or another are virginal. In the occult view, there is yet the Virgin in exaltation—the onset of maturity, of sexual ripening, desire, and most importantly, the state of all-possibility.

Purification or cleansing often presupposes a prior state of filth or defilement, but this is an imprecise and unnecessary position. All practitioners must examine their own relationship to such states as 'purity' and 'impurity' and discover whether these notions serve learning and personal evolution. All too often in such reflections, one discovers the taint of the religious, such as the Christian doctrine of Original Sin. The point is not that such concepts are useless in the work of esoteric circles, but rather that they are frequently present in the psyche of the practitioners without their knowledge, and thereby reverberate into their work without their conscious knowledge. The Pathway of the Virgin thus mandates rigorous self-examination to reveal exactly what one is composed of: points of past failure, as well as success, can be instructive in this process—if one is willing to learn.

For the present considerations of *katharsis*, let us consider the Pathway of the Virgin to be swept clean or cleared of previous spiritual states. Practically, this involves the dissolution of accreta functioning as unneeded admixtures to our communion with plants. Included among these are presumptions, suppositions, fantasies, and other psychological artifacts which often accompany exposure to that which is new and desirable. Among these too are self-importance, and the need to impose familiar structures on the unfamiliar, as well as the tendency to over-analyze. Any one of these is difficult enough to confront, let alone change: how then may all be addressed?

A valuable consideration is what one brings to the process, the Offering of Self, for this is one part of purification. In the act of the offering, the Virgin not only desires, but is desirable. In tutelary congress, there is the teacher and student; the latter must radiate desire as the furrow, the former must emanate desire as the seed. Where this mutual desire obtains, the passage of power will be accomplished to ward the singular goal of emergence. Lest we stray too far from the verdant source of our knowledge, we return to the plants themselves as teachers, and observe that all in Nature is fecundated, awash in the sexual spoor of pollen, nectar, and aroma. Such is the lay of the land, as surveyed from the Pathway of the Virgin.

Having considered the substance of the Offering of Self, and made of it the best sacrifice possible, we also consider the nature of our desire. Is knowledge sought for progression of the Pathway, or for pathological purposes? Are there emotional attachments to acquiring the knowledge? If, as a seeker after power, one can assume a state of mind in which all expectations of outcome are broken, one attains a state of placidity and pristine emptiness, likened to the Virgin, in readiness for the awakening of experience.

This 'cult of the perpetual neophyte' as I personally refer to it, is aligned in the magical orders with the grade of 0°, and is represented by the symbol of the empty vessel, or the magical circle. It assumes a constant station of receptivity toward all experience, and is assumed not only by the Virgin, but also by the Master.

Παράδοσις • Paradosis
THE PATHWAY OF TRADITION

Humanity is a matriculating species, and the biological sciences increasingly observe the teaching of distinct skills among our animal brethren. The ascent of schools of thought, and traditions of knowledge, is thus a phenomenon of our nature. Masters command their art in an exemplary fashion, and those of the correct aptitude are chosen as apprentices to pass the Art to via instruction.

Where occult herbalism is concerned, a number of pathways of tradition exist, especially in animistic cultures whose religious and magical practices have survived into the modern era. Admittedly, however, most of these traditions rigorously protect themselves from those outside the culture, for various reasons that are as valid as they are severe. Such knowledge, therefore, is admittedly not available to everyone.

However, there are also worthy traditions in education and the sciences which may be aspired to: pharmacognosy, folklore, pharmacy, medicine, chemistry, botany, perfumery, ecology, agriculture, anthropology, the culinary sciences, and theology. Occupying a middle ground between learning from a traditional healer and attaining one's M. D. are schools of herbalism, a number of which have emerged in the past fifty years. Each of these represents an empirical and academic approach to these subjects, a way that, in my own culture, is nothing if not 'traditional,' and also offers considerable breadth in the study of plants and their esoteric properties. In undertaking the Pathway of Tradition, one might, for example, decide to pursue learning the art of botanical illustration. On the surface of things, one may wonder how such an activity would lead to esoteric knowledge. Yet the creation of art by necessity involves non-ordinary states of consciousness, and for many serves as a meditation. In the formation of images through hand and eye, a certain resonance with the plant is attained.

The tradition of the passing of knowledge from master to prentice must emerge from that most rare of virtues—care. In other words, concern for the stewardship of the knowledge passed that it, like a seed, be sown in good soil. But if we liken knowledge to a seed, care and respect must also be present for the plot in which it is sown. The great institutions of higher learning do not always require this from teachers and professors, but our Way cannot flourish without it.

To walk this pathway, one must be willing to accept the rules, protocols, and decorum of the informing tradition— for tradition in fact constellates a set of rules and governances applied to its focus discipline. There will be, among some, a resistance to this. Taking occult herbalism as a point of contemplation, any rule which provokes a sense of restriction or rebellion may be considered as both a challenge to the sorcerer's level of self-control, as well as a sacrifice for the gain of knowledge. In a more immediate example, plants will impose greater rules upon the seeker than any human. Monkshood (*Aconitum napellus*) will kill if abused—that is its rule. If its rule is respected, I may safely use it to relieve the pain of a sprained ankle.

Obstacles on the Pathway are numerous, and foremost among these are maladies of the student-teacher relationship. Among these, is the ego of the teacher becoming central to the process—a problem which may stem from the master, or from the student, and often both. A strong craving for the appearance of knowledge is a pathology unfortunately all too common in esoteric circles, as is the desperate need to be 'saved' or 'given the secrets,' symptoms of shriveled spiritual virility and infantilism of the ego. At its best, the process of teaching is less about conveying facts than facilitating an experience wherein the seeker 'learns to learn' in a manner peculiarly suited to him or her. As each of us knows from personal experience, teaching is an Art which has both masters and pretenders; the true master regards instruction as a sacred process, and is willing to learn, as the Pathway of the Virgin embodies. Similarly, being a 'learner' or a student is also an Art, though it is seldom cognized as such: both teaching and learning magnify the power of this Pathway.

Another caution on the Pathway is accepting the value and limitations of anecdote. Another person's personal experience can be valuable on many levels, especially if approaching the breadth of a constant. The value of anecdotal teachings, however, must always be weighed against the voice that disseminates them, its authority, biases, and personal agenda. For example, it is easy enough to hear a teaching and accept or reject it, but less easy to ask *why* that teaching was given. Beyond a necessary engagement with basic critical thinking, common sense shall illuminate.

Aside from respect, the great responsibility for the student of Tradition is to become a House of the Ancestors. This is to say, maintaining the flame of the transmission of knowledge. In doing so, one honors one's master or teacher, and thus assures his or her place in Eternity, but also maintains a vital link in the chain that will empower future students. Such is the blessing and burden of knowledge!

Συγγένεια • Syngéneia
THE PATHWAY OF AFFINITY

In contemplating sorcery, we needs must consider the nature of that magical act of 'binding.' Most magical philosophy regards binding a spirit as commanding it, or assigning it a task or power. In most models of magic, especially those at play at the level of folk sorcery, this translates as control or coercion, with the sorcerer seeking to establish complete dominance over the spirit. With a few exceptions, this dynamic also predominates in the high magic of the European renaissance, where the operator binds or constrains spirits within a theoretical rubric of near-total control, often with the aid or permission of God and his angelic ministers. A spirit thus 'bound' passes within a sphere of magical enchantment and is alienated from its previous spirit-context; this linkage between sorcerer and *daimon* is principally defined by commandment. The coercive model of magical binding does have some applications within the realm of plant magic, but they are few, due principally to the nature of the phytosphere and the human relation to it.

And yet there is a different model of magical binding, one which resonates with occult herbalism, and animates *syngéneia*, the Pathway of Affinity. Less a matter of constraint than natural resonance, these bonds circumscribe a linkage between magician and plant based upon symbiosis or communion rather than commandment. In accord with this principle, each person—and indeed each phenomenon—despite any spiritual or magical practice, is linked to other specific phenomena purely by ipseity, or the inherent manner of its being; these strands are latent, or ambient, in that they are perpetually spun between objects and powers but not necessarily activated. In humanity, these include what we would call talents, skills, charismas, fortunes, or affinities: the natural magnetic powers of an individual that translate as powers within the world. One person may exhibit an uncanny grace and strength in feats of agility, while others may excel at high mental calculus, while others may be artistic prodigies or savants. These abilities arise naturally due to the individual's phenomenological constitution, his or her aetherically linked power, but also importantly, the magical strand or 'bond' linking them to the power in question. The strands are different for all people, and of differing qualities, but should be regarded as an essential medium for magical congress between the world of plants and humankind. As stated, many of these powers are natural, and a few can be pursued. This principle of magical affinity is not new; it follows on the ancient emanationist doctrines of rays promulgated by the Arabian philosopher Al-Kindi, and in turn upon the *vincula* ('chains'), the bonds of magic written of by the Italian philosopher Giordano Bruno.

The question for those who walk the Pathway of Affinity becomes 'what are the natural bonds between my self and plants?' This may be an affinity writ large, such as the ability to have difficult plants flourish under one's care, or it may be a kinship with a certain plant. Other affinities may lie in the sensorial realm, such as an exceedingly sensitive palate for tasting, or it may arise from the knack of being able to understand and reify metaphysical botanical arcana into physical form. An affinity may arise from strong preference—say, for a grape varietal in one's wine—that ultimately leads to refinement in understanding, and perhaps a real contribution that will progress the art form. Other affinities lie in adeptness with systems of classification, or in the art of synthesizing new combinations of individual plants. For those occult practitioners who possess Affinity with plants, it is essential to not only know specifically what they are, but also understand the nature of the linking aetheric strands, the better to pluck them, like the strings of a finely tuned instrument.

Importantly, if one has no natural affinities with plants, the sustained practice of occult herbalism—or any other discipline engaging plants—will be exceedingly difficult. This is not to say that affinity cannot be developed with dedication, or come about as the result of initiative or sudden change.

A useful practice for the exploration of these bonds begins by creating a basic map of a single known Affinity. For example, the connections present between the practitioner and the Rose, a plant perceived as common but in fact poorly understood. In the practice, one first takes stock of the sensual linkages in perception to the flower: the olfactory route, the effects on the skin, and its precise appearance in nostalgia, through personal memory. Further categories of linkage are added: links to personal history, routes of textual association, fields of unconscious association, links to dreaming, and so on. As important as any of these criteria is the field of actuation, signifying the intersection of the plant with events in one's life, especially those out of the ordinary, marking out ciphers of relation. If, after undertaking this practice, there is significant linkage or Affinity, a pattern quickly emerges, and its individual points of connection often combine to reveal a whole greater than that the sum of its parts.

Given that so many of these fields of power are personal or subjective, one may wonder what value (other than personal) such exercises hold. To that question I pose the following in reply: whose magical path do you serve? In other words, 'universal' knowledge so called is of limited value if it negates personal power: one's own personal points of engagement with plant powers are central to occult herbalism.

Another way in which to view the Pathway of Affinity is to acknowledge one's attractions, and seek them out. If, for example, one is drawn to learning about Hellebore, this should be acknowledged and acted upon, despite the plant's toxic nature. After all, the impulse toward enlightenment is unceasing, a fact too brutally illustrated in Edenic narratives. However, rather than forbidding such potent fruits, or ushering the moth toward self-immolation, we advocate the third way: the respectful approach of power. In this dynamic, understanding what lies behind the attraction is as important as the Object of Desire itself, and often serves as a counterweight to unconscious pursuit.

Περιπέτεια · Peripeteia
THE PATHWAY OF AVERSION

Everyone has things to which they are attracted; likewise there are those things which repel us. These are the things we find personally objectionable, and like those things which attract, they also contain a key to our power; their conscious route of approach is the Pathway of Aversion. For this route of occult herbalism, I have chosen the mnemonic of the Greek concept *peripeteia* or 'reversal of fortune.' This inherently presupposes the Pathway as continually active, and one which is turned into power, rather than simply opposes.

In defining that which we are averse to, we must enter the sanctum of personal abomination, a place few wish to tread, for it is easier not to acknowledge it. Yet in ignoring it, we estrange ourselves from a sanctuary of power which would otherwise control us. Aversion is one of the most difficult of pathways, in part because it involves unflinching honesty and courage. Here, it is important to remember that true personal aversion entails the things one does not like or wish to do, rather than the cultural norms one wishes to 'rebel' against.

In occult herbalism, the Pathway of Aversion often begins with the physical, and the most easily understood example is the exposure to plants causing injury or harassment. Thorns, foul stench, nauseating poison, and dermatitis are among such factors. On the Pathway of Aversion, these plants are encountered to physical detriment, and the route of recovery is sufficient to energize and catalyze new modes of magical understanding. The metaphysical aspects of the path are, of course, identifying one's demons and weaknesses and addressing them, through transforming their active qualities or extracting power from their viscera via magical operations.

Plants of the Pathway of Aversion
Blackthorn (*Prunus spinosa*)
Boxwood (*Buxus* spp.)
Hawthorn (*Crataegus* spp.)
Holly (Ilex spp.)
Locust (*Robinia* spp.)
Nettle (*Urtica* spp.)
Rose (*Rosa* spp.)
Ziziphus spina-christi

In the landscape, the Pathway of Aversion is aligned symbolically with the hedge, the division between fields, a structure of ancient origins and strong magical power. Often comprised of thorny trees and plants, such as Holly, Blackthorn, Nettle, and Crucifixion Thorn, the hedge arises as a living figuration of agricultural architecture, not merely marking a boundary, but often making it impassable, or presenting difficulty in doing so. As the terminus dividing one state from another, its power arises not only from physical division, but from the defensive power of its thorns and densely-thicketed branches. *Within* is divided from *Without*, and *Here* partitioned from *Beyond*, by way of travail and ordeal. Similar barriers exist throughout Nature, such as the airless and killing space that separates planet from planet. The Hedge thus embodies an extremity, a state of harshness opposed that which lies in its proximity.

Among some persons we observe a kind of *hedge mentality*, a state of mind which mirrors in in some ways the physicality of the actual hedge in the landscape. This mindset, often unconscious, is characterized by the imposition of barriers and perimeters at individual extremities, and the kinesthetic relationship of the Self as it moves throughout the hedge-enclosed labyrinth. Although not strictly imagined as a dense green embankment, the limits of personal comfort are strictly defined, so that there can be no personal disruption. A hedge mentality may arise from a number of legitimate personal concerns: privacy, solitude, demarcation of personal space, or the individual need for a psycho-spatial framework. It may also arise, and be sustained by paranoid states or simple denial. The Pathway of Aversion, by its nature, disrupts the ordering of hedge-mentality, and allows passage between previously partitioned fields.

Aversion may not be consciously understood, but intellectual comprehension is not necessary for the power of opposition to function on a spiritual level. So long as comprehension dwells within the body at the most basic somatic level—cells reacting to poison, and catecholamines reacting to fear—the power will enjoy a means of translation.

Αφανισμός • Aphanismos
THE PATHWAY OF DISAPPEARANCE

The Pathway of Disappearance or 'Vanishing' (*aphanismos*) may also be called the Way of the Hermit, based as it is on isolation and inhibition. That which is inhibited is human company and the comfort of the familiar world, met with a corresponding immersion into the wild. It relates to the Pathway of Peregrination, yet bases itself on removal not only from the world of humankind but also the comforts of travel. In the immersion in the wilderness, the spirit separated from the distractions and burdens of civilization, certain seekers may attain clarity of sight and occult understanding. The visionary potency of this magical process is noted by British occultist and seer Aleister Crowley in his essential work *Magic and Theory and Practice.*

There are some who argue that true hermitage in our present age is impossible, given the degree of human dependency of civilization, or the privileges of the economic classes likely to be reading these pages. After all, the posh and carefully curated 'retreat' is a major offering of the modern travel industry; wilderness survival skills courses, though appealing to a different customer base, occupy the same category. These events are not hermitage, but diversions from the everyday, designed as competitive commercial products, and are offered and mediated by others, at a price for every budget, accompanied by safety policies and written assurances of satisfaction. To be clear, there are no such attributes to the Pathway of Disappearance.

The Pathway of Vanishing or Disappearance implies dwelling within the wild, and severance from humanity without trace, as a form of ascetic practice. It is an act, and also a stance of power, needful for occult understanding. In as much as isolation and privation provides clarity by bodily removal from 'background noise,' so too does the severed bond of connectivity with all associations, and the endlessly self-proclaiming 'loudness' of humanity. This is the knowledge born of silence, a lesson taught all too effectively by Diviner's Sage (*Salvia divinorum*), whose revelatory spirit, when used traditionally, departs in the presence of the obnoxious.

This recalls the magical lesson of 'Being Inaccessible' in Carlos Castaneda's *The Journey to Ixtlan*, a message more relevant than ever in the present age of electronic interconnectedness. In cutting such ties, one understands that time, ultimately, is one's own, and how one accounts for it is ultimately a personal choice; for many, a very private one. The same applies to motion, thought, deed, and spiritual devotion.

Although the philosophers and prophets of human religions have made great use of this Pathway, the concerns of the Hermit in the Wilderness are especially appropriate to the student of occult herbalism. The diminishment of self-importance, the destruction of the ego in the name of gaining something greater. Such is the act of Vanishing to oneself, such that in the process one discovers oneself anew, the realm of magical understanding aligned with the Wildman and Wood-wife, the barbarous and atavistic beings dwelling at the margins of civilization.

Apart from the estates of soul incepted by the *absence* of humanity, we may also consider the *presence* of those powers attending on hermitage. The necessities of survival are of course paramount, incepting a shift in consciousness concerning personal responsibility, and locatedness as an organism nature. Besides direct observation of (and reliance on) the plants themselves, one awakens to unique 'languages' which interconnect flora, fauna, geology, and weather. Such are the ciphers of the Book of Nature, there to read if one dares. Within the lonely places of the wild, it is also true that dreams assume a wholly different character than that experienced previously, as the waking phenomena to which one is tethered has drastically shifted. In truth it may be said that the Path of Disappearance is such that it includes components of nearly every other pathway here mentioned, but in a particularly pure and concentrated form, the Alembic of Natura.

Yet to those who seek this way, beware the many possibilities of disruption and disheartenment. At the distant primordium of Christianity, the Egyptian hermits fled the cities seeking a closer connection to their God and his angels in the desert. In the beginning, they found demons, with whom they struggled in righteousness. In time, they discovered something resembling God. As their ascetism drew ever-greater numbers of people to the solitude of the wastes, and even became competitive—such that each sought to prove how much more righteously he suffered that his brother—they discovered demons once again.

Τοποθεσία · Topothesía
THE PATH OF WITNESS

The essence of the Path of Witness is the discipline of phenology, a term from the Greek *phainō* (to reveal or bring to light) *logos* (study), thus the study of appearances. As a pathway of Occult Herbalism its associated term is *topothesía*, meaning a locale or place, and thus we may also consider it to be the 'Pathway of Rootedness,' as its work arises from dwelling in one locale for an extended time, and incepting a certain intimacy with the powers and entities that animate its natural processes. Long practiced by farmers as a way of understanding the land, the seasons and plant and animal life, it represents a kind of 'folk empiricism' which provides precise information concerning the patterns of Nature in a given locale.

Phenology, the study of life cycles progressed over the solar year, is an essential ally of the esoteric herbalist. Whether one is a botanist, a naturalist, a healer, or a practitioner of occult arts and science, anyone can make a phenological calendar and thereby learn an immense amount of directly relevant information about plants. This practical approach can be applied to the local wild plant populations, or to those under cultivation in farm and garden, or, ideally, both. In essence, one seeks to discover the Hand of Nature by direct observation—not with a single plant, but by recording a portrait of as many as possible. This pathway demands a respectful approach, and one that may readily be educated. The collection of actual specimens is not required, save for those that shall indwell the mind and heart.

The essential work of phenology is calendrics, and the building of biological chronologies. Against a backdrop of time, plant growth and development are recorded in detail and its progression studied in relation to selected plants and their interactions with other populations. Examination of growing habit, life, reproduction, and death or dormancy is essential, noting difference between subdivisions of locality. This Pathway is of crucial importance to farmers, vintners, beekeepers, ranchers, horticulturists, and others who make their living off the land. The quality of a crop, such as wheat or grapes, is directly related to its phenologics, and assists in the husbandry of future crops and their associated products. Also subsumed within this cycle are geological processes, such as the effects of landslides and deposition, and the particulars of microclimate. The skeleton of a very simple annual approach might document the following phases of life:

Sprouting / Setting of Buds. What annual plants emerged first? Which deciduous trees broke dormancy first? Was there a new succession of weeds, in response to seasonal conditions of the past year?

Leafing. What particulars characterized the vegetative growth cycle, and were any pests present? How can the general vigor of the herb or tree be characterized in comparison to past phenological cycles?

Flowering. How early did the flowers appear this year, as opposed to the year before? Can the difference be attributed to weather patterns, plant diseases, drought, or pests? If fragrant, what do the flowers smell like, in the present moment, and last year at flowering time? What pollinators visited?

Fruiting and Harvest. Aside from such characteristics as flavor, health of fruit, and quantity of harvest, what differences can be discerned between this year's crop and the last?

The Dropping of Leaves and Dormancy. At the turn of Autumn, how quickly did the leaves turn, and how long did they stay on the branches? Was the Autumn so warm that certain annuals had an exceptionally long lifespan? In the dead of Winter, what did the bark on the deciduous trees look like, and what procession of wildlife made use of the evergreens?

The phenological calendar is often drawn in a circular form, with the seasons of the year progressing around its hub, with various concentric rings representing various species. Simple individual models can be made to accommodate various grouping of plants such as fruit trees, annual medicinals, grain crops, or cut flowers. The visual display of information should be made in the manner most suiting the practitioner, and in terms of the information documented, there is no limit save the boundaries of the imagination. I have kept phenological calendars of plum trees of different varieties, and how their life cycles over time are related to wine made from their fruits.

The Path of Witness may, upon first consideration, seem quite the opposite of what is understood to be 'occult' discipline, but let us consider that the information thus collected is ignored by most, and thus is 'hidden.' It is, therefore, a kind of concealed knowledge readily offered in generosity by the Hand of Natura to those who would listen. Consider too that gardening books and botanical manuals give only general information about how and where an individual plant species plant might grow: the most accurate information to be gained for the individual practitioner is to be gained by watching what is revealed in the land of their own tending. The more one pays attention through the keeping of such records, the greater the secrets that shall be revealed; one should not be surprised to discover properties of plants that contradict both scientific and occult orthodoxy. This revelation of the nature of the patterns of place is in fact a partial portrait of the *genius loci*, or spirit of place.

The nature of phenology is inherently pure—it cannot be capitalized upon in a broad way, because it deals inherently with locality, and thereby intimacy: as an art it renders up secrets that will be of direct benefit to the scribe. Its greatest importance is thus to the practitioner who creates it, the land of its immediate documentation, and the local community of plants, animal, and mankind.

Περιπλάνηση • Periplánisi
THE PATH OF PEREGRINATION

The Path of Peregrination, or Wandering, involves physical movement between locations, both physical and spiritual, for the gaining of magical knowledge. It may be considered opposite, yet also complimentary, to the Path of Witness. Its power lies in diversity of locality, and in the gnosis gained by movement. This way of learning arises from the exposure to that which one has never seen, but also gaining understanding of the locations themselves, and how plant power manifests there.

It is a fact of botany that plants vary in their attributes by location, and there are countless examples of this. A conifer growing at the timberline may be stunted to an almost impossibly small size; the same tree at lower elevations will be tall and stately, though they are the same species. The stately California Live Oak (*Quercus agrifolia*), usually a tree of grand bearing, becomes stunted and wizened along the central coast fog belts, earning the name 'pygmy oak.' A medicinal plant, such as *Salvia divinorum* may have differing concentrations of medicinal principles based on whether it grows in a sunny location or a shady one. The Pathway of Peregrination allows the seeker perspective on the diversity of plant species, affording a kind of intimacy that textual education cannot.

Similarly, the traditions of plant magic and medicine will vary by location, in accord with the history of the human culture situated there. The solanaceous plant Pitchuri (*Duboisia*

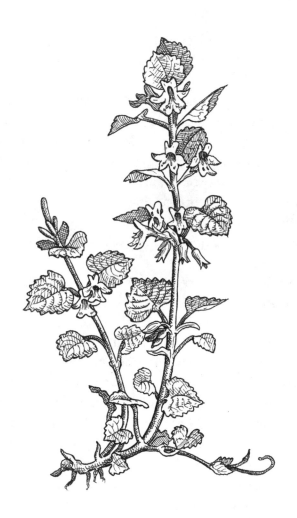

hopwoodii) used in its Indigenous Australian context, will draw directly from its reservoir of power in the land and the people, while the same plant grown in a botanical garden in England will not. Likewise, one plant may be used for completely different magical purposes, sometimes as divergent as spells of healing and cursing, in geographically proximal areas. This phenomenon is one that has yet to be addressed in books of magical plant correspondences, as the human mind is trained to seek simple, singular answers, rather than complexities. The Path of Wandering thus affords us the opportunity to transcend simplistic models of understanding plants and their power, forsaking the insular world of the library for the real world.

This leads us to the study of plant communities, a subject of incredible importance, not only to ecology, the discipline whence it emerged, but also to esoteric botany. Each plant not only has favored conditions for growth, such as soil type, climate, and amount of light, but also those plants it prefers to grow near. These botanical alliances may serve, through the doctrine of correspondences, to indicate unusual spirit-presences and/or to potentiate the powers of each plant in the community. Illustrative of this concept is a place I have often walked to, its sole plant inhabitants are Blue Elder (*Sambucus cerulea*), Poison Hemlock (*Conium maculatum*), Stinging Nettle (*Urtica dioica*) and Poison Oak (*Toxicodendron diversilobum*). Such an aggregation of poisonous, witching and 'hostile' species serves not only as a shrine of contemplations of the great Circe and her powers, it also affords privacy against the intrusions of humanity.

As magical practice, the Pathway of Peregrination offers many extrapolations, but perhaps the strongest of these is a plant pilgrimage—a mindful journey specifically devoted to the purpose of seeking a specific plant, or learning about the herbs of a certain area. In this manner, the Pathway has offered up precious revelations that my home locality could not, such as fern communities thriving in volcanic steam vents, or the sacred Egyptian Blue Lotus (*Nymphaea cerulea*) growing in ditches on remote islands of the South Pacific.

Importantly, the power of the Pathway of Peregrination arises not only from the loci themselves but from the interstitial process of movement between points, catalyzing new understanding. *In situ*, we may partake of power in its concentrated form; in motion, as we journey between points on the map, that which has entered our consciousness becomes reflective and connective, tessellating itself with our own knowledge base.

Επιστασία • Epistasía
THE PATHWAY OF THE STEWARD

Where harvested products are concerned, stewardship of plants provides assurance of quality, as well as a personal and living connection to them. We may contrast this with the impersonal nature of plant obtained by commerce. It may also be called the way of *Therismos*: The Path of Reaping and Harvest. This direct connection by working on garden and farm allows for direct learning based on close observation, and trial and error.

The farm or garden is an immense crucible of instruction for botanical knowledge, providing a way of life highly attuned to the earth and the seasons. Amongst its many branches of art are the knowledge of microclimates and soil, plant economics, exposure to plant poisons, propagation, pests, plant ecology, harvest and farming techniques, and more. We are also offered the introspective knowledge gained in the post-harvest activities of vintning, distillation, canning, pruning, crop rotation, and many other skills. More than this, the farm and garden is, in itself, a highly concentrated metaphor for the tending and progress of the human soul.

Having worked on farms, and in gardens, landscapes, and greenhouses, I have come to understand a wealth of things about plants no human or book has taught me. For example, which parts of the Cedar tree are the best for incense, the fact that Poison Oak may attain woody trunks like trees, and the peculiarities of altered consciousness brought on by inadvertent dermal poisoning by hand-pruning a large Oleander hedge. Aside from practical matters of sowing and reaping, one gains access to hallowed plots of such agricultural deities as Saturn, Ceres, Priapus, and Vertumnus, and experiences in a direct manner their patronage.

Epistasía, the Way of the Steward, is above all about protection and care, and the process of empowerment it fosters. An elementary model for practice is the simple garden kept under one's own care, with plants curated according to need, affinity, fascination, and suitability for climate and soil. Special consideration should be given rare or endangered plants, or heirloom varieties with sound genetic lines. Over time, successive generations may be observed as the contours of the garden expand, and each crop shall give forth its instruction to the ears of the attentive.

Σύνθεσις • Synthesis
THE PATHWAY OF COMBINATION

The Pathway of Combination, or Synthesis, is the approach of combining powers, plants, disciplines, and practices to achieve a desired amalgam. Its approach, which in part is rational, is characterized by sorting, extraction, analysis, blending, and testing, the better to understand the whole and its fractions. It chief pursuit is that of the Artist, who chooses colors from the palette and conceives a world in graphical form. In esoteric traditions it is one of the primary engines of the alchemical process; in the everyday world it drives industry. Synthesis mediates the Pathway of the Virgin, which concerns all-potentiality, and the Whore, which embraces all.

A longstanding argument within occult circles concerns a taboo on 'mixing the planes,' in other words, avoiding combinations of elements of different religious or magical practices within the same ceremonial operation. Colorful tales are told of instances where this was done and great woe resulted. However, in the scope of history, Religious Syncretism is an example of this principle in action, and there is no doubt that in a great many cases of this the spirits and gods have attended and animated the co-mingling. Christianity itself is replete with such recombinant spirits, as attested to by the vast number of saints and shrines whose attributes were formerly those of pre-Christian gods. If one accepts that 'the planes' exist, it follows that they are subject to their own energetic parameters, and will react accordingly if co-mingled to the point of abuse. The admonition then, might be more intelligently applied to careless appropriation, a disrespectful and craven impulse, and one that most often results only in a surface parody of the appropriated power.

The genius and value of synthesis lies in its ability to reveal new powers from previously unknown combinations, a modality most useful to the occult study of herbs. A practical application of the Pathway of Synthesis may be readily understood in the context of healing: the study of drug combinations and interactions, a skill of absolute necessity for the clinician. Such also involves the art of Diagnosis— assessment of the patient, and connecting him or her to the correct treatment. In such modes of synthesis the virtue of discernment is as important as the artistry involved in preparing the treatment. As an active parallel, certain plant powers are strengthened or weakened when in the presence of other herbs. Understanding of this principle unlocked the great corpus of *pharmahuasca* sacraments, the Ayahuasca analogues created by combining psychoactive tryptamines with beta-carbolines.

The first step on the Pathway of Synthesis is imagination, in which possibilities are conceived by the unfettered visual apparatus of mind and spirit. A plant may directly be the source of inspiration in many cases, a viridian Muse who provides the liberating substrate. This primordial emanation, by its nature, dreams of what is possible, based in part on actual knowledge, and also in fantasy. The seed of this imaginal process is thus captured as a theoretical model and, returning to ground state, tested in the realm of physical phenomena, drawing ever closer to the crystallization of the Stone.

The Way of Synthesis is not only applied to plants and their combinations of powers, but also to the individual path of the seeker. Like the human body, the spirit is sublimely adapted for synthesis, and the attentive will note that the most potent and effective teachings, those which endure over time, have the character of continual revelation, even if originating from different sources, or separated by great lengths of time. More than this, such teachings mutually support and recombine with one another, generating a state of recursive gnosis that is in itself a 'divine mixture' uniquely formulated for the Self.

Συνομοσπονδία · Synomospondía
THE PATHWAY OF CONFEDERATION

In past writings[1] I have referred to 'the Emerald Convivium.' This term is meant to comport alliance between individuals who study and practice occult herbalism. It is not an actual organization but rather an inherent spiritual affiliation between those who study plant powers. As the Pathway of Confederation, it may also be understood as an active approach to learning, characterized by the gathering of kindred learners, and learning through research as a collective.

Plant traditions have long histories of such organized groups, devoted to such things as seed saving and exchange, and the modern incarnation of community-supported agriculture. In occult herbalism, the Pathway of Confederation suggests the formation of independent occult study groups, organized around a mutually decided sphere of investigation. Models for such might include the study of a singular plant for different periods of time, the investigation of certain authors or books, or practical comparisons between different plants.

---

1 *Viridarium Umbris: The Pleasure-Garden of Shadow*. Xoanon Limited, 2005.

As a simple example, this latter category of assembly might take the form of an incense or essential oil group, where concentrated expressions of several different plant products are sampled by attendees and discussed. This model presently operates in private wine clubs and whiskey societies which hold regular tastings, each event organized around a particular datum such as products of a single distillery over many years, or in the case of wine, vertical or horizontal tastings. This approach provides a ready avenue of direct experience, and in the best of cases will present opportunities for affiliation and collaboration with like-minded practitioners. Other concerns for such a confederation might include:

- Work to recreate lost occult recipes or sacraments.

- Excursions or 'field trips' to the wild or to public gardens or arboreta, for the purposes of familiarizing oneself with certain plants.

- Gathering data and actively teaching specific botanical processes, such as the making of wines or perfumes.

- A group based on the herbal doctrines of planetary correspondence, examining the different models that have been propounded over time, and their usefulness.

- Experimentation with certain magical operations, such that each group member performs the rite in solitude, then reports back, and results are compared.

- Forming a consortium with the purpose of protecting access to rare plants needful for ritual practices. This concerns economics and supply routes, and has long been an essential consideration for all ceremonial organizations, whether magical or religious.

- A group structure built on private meetings, but holding occasional public events, such as a fair, colloquium, or sale.

As is readily apparent, the cartulary of possible foci for such a confederation is literally endless. Choosing a focus for such a group is the simple part; the more difficult parts arise in the arts of diplomacy, generosity, collaboration, and organizational acumen necessary for maintaining a functioning group. Some may even arise organically, cohered solely by the force of mutual fascination. Having participated in many such groups, I have found the successes and failures of each to be instructive. The constant of every such organization is its animating genius, which, if it remains vital, is sufficient lifeblood for its sustenance.

The difficulty of course with such groups is also the same as their benefit: people. This may be a matter of personal temperament: some simply do not wish association, preferring instead to work alone, thus the Pathway of Confederation is contraindicated. For those who do prefer working in groups, there is the perennial plague of certain individuals who seek to make the collective a medium for the inflation of their egos, through causing discord. Still other instances will arise when one becomes a member of a group only to find it does not personally resonate. In such cases, if the group does not serve the personal needs of the individual, one need not attempt to bend it to one's will or to make an obscene display of emotions—one only need simply leave. Such is just as potent an exercise of Will as joining in the first place. In short, if one is committed to occult learning in a group context, one must understand the conditions necessary for such to thrive, and incept standards to protect them, as well as create and enforce personal boundaries against the inappropriate.

Above all, the glory of the Pathway of Confederation is that one's own manner of perception is enhanced, challenged, and affirmed by others, and that multiple points of learning and inquiry potentiate the arena of learning. This happens in the best of groups, where principles of affinity and opposition are both present, but serve the central aim. Although it should not expected, lasting alliances can result from such crucibles of teaching, not the least of whom are plants.

Ἀποστασία • Apostasia
THE PATH OF TURNING AWAY

There comes a point in one's spiritual path when one must turn aside from all books and teachers. In our rubric of the Pathways of Occult Herbalism, we call this the Pathway of Apostasy. It is related to the Pathway of Aversion, in that there is a dynamic of conflict yet its power is more centered in Opposition, and the turning against what one has known. One may consider its essence to be rejection or disavowal, instead of the confrontation that the Pathway of Aversion implies.

An immediate association with such acts is the 'turning away' from establishment religions that often serves as the catalyzing impulse for the spiritual seeker. In this action, there is much to disavow, from proscriptions on thought and behavior to doctrines which impede the progress of the human spirit or are unconditionally destructive to the world. Yet the most potent and appropriate impulse for turning away from such religions is their failure to resonate with the intelligent and empowered individual, an offense to the personal center of gravity.

Turning from that which does not serve the seeker is easy, at least in theory. Bonds of attachment have already loosened, if not broken, due to lack of personal connectivity. But can one turn away from that which *does* serve, specifically to harness the power liberated by the turning? Such is the dual way of Truth-in-Treachery, a philosophy embedded in the stream of magic known as Crooked Path Sorcery, an ethos of the Sabbatic Tradition of Witchcraft.[2] The conscious reversal of positions, sometimes by the most severe of ritual ordeals, is within this tradition often referred to as 'dancing upon the edge of the Blade of Cain.'

However the blade of heresy cuts both ways, and may turn inward: in the acquisition of knowledge through the Way of Apostasy, be prepared to have all previous experience challenged, and the ego tempted with false knowledge. Be prepared to learn more than you wished to, including your weaknesses, inadequacies, and psychopathologies. 'To what end?' ask the throng of sybarites. Such is the occult nature of medicine, and its aim.

In occult herbalism, as a philosophical exercise, a temporal state of apostasy may be induced by reflecting upon one's learning and asking the question, 'If not…what then?' In the mind, all previous teachings—and in the purest forms, all attachment to gaining knowledge itself—are thus immolated as sacrifice. Within the ashes of this pyre, new understand-

2 See Andrew Chumbley, *Qutub* and *The Dragon Book of Essex* (Xoanon, 2005 and 2014 respectively). The book *Via Tortuosa* (Daniel A. Schulke and Robert Fitzgerald, Xoanon, 2017) a book concerning the metaphysics of Crooked Path Sorcery, is forthcoming.

ings arise. Many are but the nascent glimmerings of what may be, thus the Way of Apostasis relies also on intuition, due to its rejection of other guiding compasses.

While any plant is capable of causing an overturning or re-ordering of personal knowledge, certain individuals may be considered 'patron saints' of the Pathway of Apostasis. These include psychoactive plants with strong dissociative tendencies such as Thorn-Apple, as well as Ayahuasca and its retinue of plant spirits. In North America, a traditional example of this profound 'turning away' occurred in the Algonquin manhood ceremony of the Huskanaw, during which boys were administered the *Datura*-containing potion *wysoccan*, which, through its strong dosage and dissociative action, induced a state of permanent forgetting of their former lives.

Ακολασία · Akolasía
THE PATH OF THE WHORE

The Pathway of the Whore is like that of the Virgin, in that the giving over of oneself to congress with the Other is an act of power essential to the route. In the case of the Whore, this act of self-sacrifice is characterized by its detractors as 'indiscriminate' and thus perpetually contaminated; and by its proponents as 'all-inclusive' and thus possessing the ability to contain entirety. Ironically, this philosophical position emerges from an attempt to direct a power that resists control, and endlessly self-pleasures in the face of all moralities. Its contemplative word is *Akolasía*, Greek for 'promiscuity,' 'wantonness' and 'debauchery.' In essence, it invokes the path of excess, the precipitous route of adopting a great many possible routes, and with enthusiasm.

In my own history of learning the ways of plants, I have often observed the pathway of the Whore as an impulse, barely conscious, of the person newly awakened to the plant world, and to their own passion for learning the secrets of botany. There is an almost uncontrollable rush to assimilate as much knowledge as possible, and in the midst of such enthusiasm, subtlety and nuance are lost. This changes when the path is conscious, for with every extension of oneself toward embrace, power is sent forth, and the present knowledge and application of will toward this act defines sorcery.

As with other Pathways, there is danger: besides death, addiction, demonic obsession, and disease, there are other perils more subtle, particularly for the esotericist. Overexposure to one plant ally can, for example, severely affect interaction with others, distorting or suppressing certain organs of sensation that they become effectively useless; Cinnamon is one obvious example, as are Opium and Tobacco. Even in such cases, the knowledge gained by the practitioner once ground state is restored is invaluable.

The Pathway of the Whore, then, in order to gain the most power from interacting with the spirit world, must learn the practice of *fractionation of Self*, the better to offer oneself in carnal embrace, and receive the embrace of the Beloved. This may be understood to be a willful state of dissociation, wherein one willfully fashions a number of vessels of psyche, each to fill with an id-entity suitable for the congress, thus partitioning, or compartmentalizing the fractions of psyche. Forms of this practice are well known to sex workers, as well as actors and field operatives in espionage.

To understand how the fractionation of Self applies to the Arcanum of Excess, one must also know Temperance, or reserve. As specifically applied to sexuality, this is encapsulated by the concept of chastity or sexual continence, ideas which, in a general sense, are heavy laden with opprobrium. However, control of the sexual drive is a neutral concept, and a time-honored technique of occult practitioners. There are a multitude of techniques to develop this, but the practice of *karezza*—the suppression of climax after being sexually aroused to near-orgasm—is exemplary. In a more psychological sense, one can immerse oneself deliberately in situations which are sexually arousing, and, harnessing the energy thus generated, direct it into a non-sexual channel. The somatic and psychic power thus raised becomes undifferentiated and freed from previous attachment, able to be used according to will. The ultimate aim of such work is not abstinence or prudery, but rather the conscious calibration of the sexuality of the practitioner, much as a rheostat, such that one apprehends the full range of control over the current, as well as the implications and necessities of losing control at the appropriate moment.

Similar benefits of Temperance are also true in Excess. A perfect example is the knowledge gained by overdose, particularly in cases where such symptoms deprive a person of their usual sense of control. The visionary Solanaceae or Nightshades are such plants, toxic to humans, and are characterized in overdose by extreme psychosis, dissociation, and alienation from the ego. The relinquishing of personal control to a plant entity, one that can be deleterious to one's very incarnation, represents an expansion of the Pathway beyond its central track and into its frontiers: the ditches, brush, and badlands of the Way. Whether this is characterized as careening off the path or as an instructive diversion is largely determined by whether one has a body to return to afterward, as well as one's ability to take a hint. Yet those who survive, foolish or wise, will understand that whoredom does have a circumference, and this knowledge will perhaps temper future licentiousness.

Another difference in approach between that of the Virgin and the Whore is the latter's greater knowledge of the body, an evolutionary principle of knowledge sometimes characterized by the vulgar phrase 'use it or lose it.' This is the difference between highly evolved, specialized organs and those which over evolutionary time wither and become vestigial: practice makes perfect. Having thus attained the Wisdom of the Body, the practitioner is aptly prepared for the Thirteenth Pathway.

Ενσωμάτωσησ • Ensomátosis
THE PATH OF EMBODIMENT

All human experience is of the body, whether acknowledged or not: learning occurs through the numerous gateways of the mortal flesh. This primal means of ingress serves as the basis of the Path of Embodiment, encompassing the respective somatic processes of incarnation, experience, incorporation, synthesis and becoming. The understanding accumulated in this byway is known as *corporeal knowledge*—knowledge of, or from, the body. Though distinct, it is the unifying principle of the previously stated pathways, purely by virtue of its permeability and the degree of intimacy between the human *corpus* and the plant world.

In approach, this would seem a simple route, given what has come before. You, reader, are a living construction of blood, bone, and tissue, a fact witnessed by the hand that holds this book and the eyes that read its pages. It would stand to reason that the Pathway of Embodiment would be the most elementary of routes to acquiring knowledge of our Art.

This is a fallacy, for the fact of inhabiting a body does not assume that this act is conscious, nor that it is properly cared for or exalted, nor that it assumes its greatest potentials, be they magical, or otherwise. In cultures which stridently deny the body—a putrefying legacy of religion—a series of protocols are imposed and adopted as second nature which either restrict, or in many cases actively degrade and destroy, the physical vessel. These may range from sedentary activities to unhealthy environments or dietary habits, to psychological derangements based on demonization of the body. In other cases, such corporeal impediments are deeply cherished, such as rational thought, the cleavage to which often obfuscates, minimizes, or denies the experience of the senses, seeking to explain experience purely from its own narrow purview. The unbinding of such restrictions, and the awakening to a consciousness of the flesh, will naturally result in the sudden awareness of one's own state of corporeal unknowing. Yet the sooner one discovers the expanse of one's own ignorance, the more quickly one realizes the vistas of opportunity for learning. This is especially important especially as pertains to the magical arcana of plants.

To walk this pathway, it is also necessary to commit one's body as a laboratory of learning, a conservatory of the instruction of herbs and trees. Reliance on one's own corporeal experiences for active instruction, particularly as concerns the individual's spiritual path, is an essential part of this, but one must also learn to accept oneself as a valid somatic gateway for information. This assumes, of course, that one is not a liar, or particularly prone to denial, for self-delusion is a great enemy of experiential knowledge, and where occult herbalism is concerned, certain experiences will erect the boundary of fatality or maiming against pretenders.

To embody a plant mystery is to become it, to 'make flesh' of its teachings by living them as reality. We observe that certain plants assimilate and emanate specific powers, not only from the symbolism that humans have applied to them, but through their behavior, morphology, ecology, and chemistry:

Apple: Beauty, Wisdom, Enticement
Datura: Seduction, Revelation, Phantasmagoria
Dodder: Parasitism, Encroachment, Compulsion
Euphorbium: Antagonism, Resistance, Corrosiveness
Nepenthes: Glamour, Ingenuity, Monstrosity
Oak: Strength, Nobility, Sustenance
Poppy: Mercy, Tranquility, Enslavement
Rose: Nobility, Love, Sacrality
Vanilla: Allure, Promiscuity, Arousal

Through the Pathway of Embodiment, the herbalist understands that each plant represents attributes or powers to which he or she may be apprenticed, and thereby assimilate, a kind of totemism. When such virtues have become second nature, to the point that the practitioner is empowered, one may be said to have embodied the attribute of the plant. Importantly, such embodiment entails a holistic embrace of the entire corpus, such that all of homeostasis is affected: it is simply lived, rather than spoken of.

An exemplary practice of the Pathway of Embodiment involves a form of the *Hieros-Gamos* or sacred marriage, and demands the absolute commitment to a single plant for one full cycle of seasons. Old rites of marriage between a human and a tree, or between a monarch and the land, carry the eldritch concentrations of this mutually interactive power. The methods by which this is done must be private, as one is essentially giving one's substance, through the crucible of ritual, to Nature. In the simplest of terms, such an act is a promise made with perfect sincerity that one will put aside ego and ready oneself as a vessel of apprenticeship.

For the length of one year, a single herb, or plant, forms the solitary focus of one's work, contemplation, and ritual practice. The herb is chosen based on its harmony with the human body, being sufficiently non-toxic to a degree that the plant may be eaten and drunk daily. The practice begins with prayer and offering directed to the plant itself, in its living form, seeking to learn what one may through the intercession of the herb itself. To the greatest extent possible, the plant is incorporated into the life of the practitioner, through consumption of its leaves, roots, flowers, fruit, and seed, and through feeding the plant substances such as blood and hair from the practitioner's body. Throughout the year, not only is the plant consumed and drunk, but also grown in the garden, or visited in the wild. It should be cared for and given offerings, but also form the central focus of mystical contemplation in the form of a shrine, whether real or imaginal. The practice is rigorous, and, to the reckoning of some, ascetic, but worthy. At its conclusion, one may not have precisely embodied the Arcanum of the plant in question, but without question, one will have achieved a certain kind of expertise, intimate beyond book-learning, about the herb or tree.

The Gardens

Turning from the Ways of our approach to the destinations themselves, we now consider occult herbalism as a series of enclosed gardens, each a crucible of concentrated *viriditas* animated by a different aspect of the Green Art. I have here numbered them as thirteen, the better to achieve simple numeric harmony with the Pathways we have examined, but in truth, the arcana of trees and herbs are uncountable, their revelation limited only by the hardened carapace of the intractable heart.

Each garden is, as a pleasant exercise of the mind, conjured within the imagination, the better to understand its *numen* and substance. Thus fixed within thought, the particulars of each may go forth into the sensorium, blossoming as a newly

opened bud, unfolding as mental devices of imaginal concep-
tion, ensouled ciphers, or occult architectures which may be
entered and explored, yet leading on to others unwritten.

In the case of each, one beholds a pleasant sanctuary with
fountains, pleasant trails, follies, and an arrangement of trees
and plants which best suit the focus of its governing discipline.
The geometry of each garden is as important as the spaces it
encloses, for the arrangement of plants and their powers in
relation to one another not only affects movement through
the garden, it also embodies certain occult principles that
define each enclosure. As one is immersed within each green
locus or *viridarium*, its arrangement of plants, powers, concepts, landforms, and *animae* will ignite understanding, and
thus constellate in a tutelary flood its zone of occult botany.

The Garden of Ecstasis

At the great gates of the Garden of Ecstasis, one is met by
the robed and masked Granter of Permissions, greeting each
wayfarer with the challenge, *'What wilt thou?'* Each delight in
this garden is arranged to appeal to the pleasures of the flesh,
not as vices so called, but each sensation as a potentiator of
ecstasy all its own. At the center stands an Apple orchard of
the most ancient and rare stocks, with fruits so diverse and
astonishing that their mere aroma seems to transport one in
body to dwell among the fair-limbed Hesperides. Behold too
the many plots devoted to Viticulture: here are vineyards with
the most ancient stocks, planted by Dionysos himself, of va-
rietals unknown to mankind outside these walls. The dwell-
ers here undertake the study of the governance of *terroir*, with
wine-presses, vast cellars, and pleasure-houses for drinking

and feasting. An entire school within this garden is devoted to the ancient art of mixing inebriating herbs with wine, and the locus of the Amatorium is planted with Myrtles, Artemisias, Satyrion, and other aphrodisiac simples. Beyond it lies the great meadows of the Orgia, where all here dance and cavort by night. Sprawling culinary herb gardens lend their flavors to breads, pastries, liqueurs and other delights. Ever a place of perfumed delight, it is also a bright and leafy place alive with music, for a great number of herbs and trees which grow here provide the sounds of the drum, syrinx, and lute.

In exchange for convenience, speed of acquisition, and the lessening of physical labor, urban humanity has experienced a great benumbing of the flesh. Exacerbated by the tedium of religions preaching anti-sensualism, punitive governmental regulations, and scientific institutions plagued by prostitution to industry, the senses yearn for that which lies beyond, or that which may simply be experienced freely. Lest we despair, commerce stands ever at the ready to introduce new taste sensations to the bored palate for a price, each more perfunctory than the last, complete with accompanying platitudes about exactly how one should enjoy one's glass of wine, and why drinking the bottle on offer makes one more sophisticated. You may even be given an explanation, in advance, of exactly what you should be tasting. It is perfectly natural then that modern city-dwellers seek expansion of the sensorium through ecstatic drugs, divergent sexual techniques, and occult practice, for the human organism is evolved for sensation, and ever seeks it out. When neglected, or suppressed by a mediating force, the senses can become atrophied, just as an unused muscle. In this estate of body, the study and practice of the occult arts is compromised; the Garden of Ectsasis thus takes as its sacred focus the plants that arouse the senses and all their attending rituals, that the worlds in *all* their forms may be more thoroughly embraced.

The Garden of the Sepulcher

A vast garden, it is prepared on the venerable plan of the necropolis, with tombs, crypts, and monuments of white and gray marble forming its hardscaped foundation. Amid the tombs stand arboreal specimens of Cypress, Yew, Cedar, and Arborvitae, the immortal soul-trees. Beneath their branches the pilgrim of this garden may sleep, and perhaps dream of them who have passed into shade. In wandering the garden, one may process from shrine to shrine, each venerating a beloved ancestor, and perhaps to the altars of the death-gods themselves, there to burn incense of Storax, Frankincense, and Sandarac. Hidden away amid the trees is the undertaker's chapel, together with the woodshop of the coffin-maker and totem-carver, the implementarium of the grave-digger and those who cut sweet-smelling woods for the funeral pyre. As befits its somber purpose, a great portion of this garden is darkened in shade; yet high on an arid hill, towering above the necropolis, stands of Myrrh trees flourish, queen over all which lie below.

The selpuchral garden concerns the preparation of dead bodies, embalming, funerary plants and their ritual, and the ongoing cult of the ancestors. Here too one may learn and study the myths and realities of the suicide-potion, and of the black arts of necromancy, for each of these disciplines has an attending retinue of plants that mankind has used from the most ancient of days. Yet, in the herb-lore of old, as well as in funerary magic, the gardens of mortal death also contain within them the powers of resurrection, and thus a nuanced understanding of life. This is only paradox in seeming, for it is difficult for some modern minds to conceive entire societies, philosophies, arts, and religions thriving on a spiritual foundation of metempsychosis. Nonetheless, the herbs show us the way, in their emergence from dormancy, in vegetative re-growth from felled trees, and in the enduring traits passed from parent to offspring via the seed.

The Garden of Adornment

As a monument for the daughters of primordial *Terra*, the Garden of Adornment is replete with statuary, pleasant bowers, and pools for bathing. In this garden, wells and mirrors abound, both for the vanity of Narcissus and the beauty of Venus—for each wields an awful power. Olive and Almond trees, providing rich oils for the skin, dominate the plantings, as do preferred cultivars of Apothecary Rose, Jasmine, and Lavender. Here grows azure-budded Ceanothus and all saponified herbs, for the making of soaps and lotions. In a shrine all her own are reverent plantings of Belladonna, the beautiful lady, her dark berries to stain the lips and open the mesmerizing eye. A heady fragrance arises from the Acres of the Virtuous, where grow sacred flowers nourishing beehives, whose honey and wax yield balms and depilatory creams, for smoothing the skin. In the Garden's dark and wooded depths, hidden from the eyes of many, stands a round stone chapel whose frieze is graven with the images of the fallen angels, each with his earthly bride. Its interior conceals a deep mine shaft whose interior walls bear crystals of malachite, lapis lazuli, kohl, red ochre, and all those minerals harvested from the womb of the earth for painting the face.

In occult tradition, the Garden of Adorments is *kosmesis*: the ornamentation and enhancement of the living body, and nurturance of the powers of beauty and physical charm. As with Alchemy, cosmetics are of ancient origin and make use of plant, mineral and animal substances. Plants so used may stain the lips, dye the hair, tonify or depilate the skin, sweeten the breath, or brighten the eyes. This sublimely powerful garden, governed by Venus, returns us to two ancient magical concepts: that of the glamour, and also the mask. Glamour is the practice of affecting the world through the power of projected allure, specifically through the appearance of the face and the body, as well as the related characteristics of movement, manner, and posture. This is a recurring feature of ancient witchcraft tales, wherein decrepit bodies are temporarily restored to sleek, taut, and comely youth, the better to seduce the victim. The mask is the construction of an intermediary form of the Self, either in material or spiritual form, a temporary ego-concretization that concentrates and projects certain powers, while suppressing others. Both of these powers belong to the shaman, and form a portion of the sorcery of shape-shifting; although each has a material aspect, its hidden *anima* is much greater. Cosmetics, seldom considered a worthy topic of occult discourse, is nevertheless placed historically at the very beginning of magical time, when it was taught by the fallen angel Azazel to human women, one of many such 'forbidden' arts that served as the primordial foundation of occult sciences.

The Garden of Zôion

Vast in scope and botanical diversity is this garden, at once a savannah, a rain forest, a range, a paddock, and a carefully curated physic garden, bound as one. The secret courtyards of these expansive grounds are planted with Angelica, Violet, Valerian, and those talismanic simples used for the magical control of animals. Growing hard by, one may wander amid well-kept Yew coppices for making the longbow, and also discover the Acacia with its punishing thorns, ready material for making fishing hooks. Plants used as fish poisons, such as Tephrosia and Trichostema, line the banks of a flowing stream, where bending Willows are used for making traps and nets. In this garden one also finds vast apiaries, and fields of wildflower blossom for forage, and ranges of grasses for pasture. The trails meandering through this artificial wilderness belong as much to the deer as to the human. This garden, though ordered, also embraces a portion of land left wholly in its pristine wild state, tended by the Hand of Nature.

The Garden of Zôion encloses the zone of occult herbalism specifically relating to the magical affinities between plants and animals. Its allied concerns are herbal veterinary medicine, hunting, and animal husbandry. With regard to hunting, some of the oldest and most widespread forms of plant magic utilize plants as magical charms for success in hunting, to lure quarry, to bless the weapons, or to propitiate spirits of the hunt for good luck. Also to be found here is the ancient form of magic known as beast-charming, which serves as a means of magical control over animals, using the mind, word, gesture, and of course plants.

The Garden of Hekat

Warmed by steam vents, the Garden of Hekat recalls to mind the environs of Avernus, its mist and moisture to better nurture all the venomous luminaries of the phytosphere. Its entrance is framed by a grand arch of deep virid Serpentine, carved with a representation of Mithridates VI Eupator, his noble form reclining on a bed of skulls, and a great procession of living herbs proceeding from his outstretched hand. As one enters the garden, one is suddenly overcome with a sudden effusion of musty and acrid fumes: these arise from dedicated beds of poisonous umbellifers, where grow Aethusa, Cicuta, Conium, and Venomous Angelica, as well as Colossal Hogweed. Fallen among them are broken statues of the many generals, politicians, and rivals they have annihilated over the centuries. In the western quarter of the garden, a stone pavilion stands amid a vast planting of Datura varieties: within it are opulent beds where the visitor may sleep at night, his soul borne aloft on the bewitching fragrances of the Devil's Trumpet. The dream so given is said to originate in one of two places: from the licentious ravishers of flesh known as *incubi* and *succubi*, or from Death herself. When one is awakened, the dream persists in the memories of the body, uniting the sleeper with the denizens of the garden. In subterranean chambers where disease and blight are nurtured by the gardeners, Ergot flourishes, as do charms of crop-withering, and in the root-choked darkness the fatal beds of Toxicoscordion yield up their abominable tubers. Shrines to the flayed gods rise from sprawling patches of Metopium, Toxicodendron, and other plants which burn and blister the skin. By night, the narrow paths are lit by the Mandrakes Illuminant, sending forth rays which first set the mind to sleeping, then awaken it to Satanic power.

Poison: its essence is that which causes harm or disruption to the body, whether in the form of death, dysfunction, maiming, or a sudden and violent catapulting into the realm of an alienated Self, called by some 'dissociative psychoactivity.' Its flowering is the Greek *pharmakos*: magic, medicine, and poison, all of which are assimilated in lore to witchcraft. Another great legacy of folk herbalism is in plant-derived pesticides, used for repelling, stunning, or annihilating vermin of various kinds. This has an occult corollary—the repelling of demons, and the sending of curses upon enemies. Their occult power notwithstanding, the witching brews and salves are but a part of the mysteries of Poison, the others being Antidote, Intoxication, and that middle ground between fatality and therapy, Anesthesia.

The Garden of Perfumery

In its sun-illumined plots, the best roses and gum-bearing trees creating stylized floral patterns to amuse and delight the wanderer; the air is thick with spices and the rising sweat of the Storax, the enticing floral charms of Rock Rose and Neroli. Throughout the garden, pools and stone walls have been carefully constructed, the better to channel air currents in different directions, churning up rivers of scent that one may distinctly imbibe as one walks. A temple of Venus, she of allure, forms an important shrine near the great distillation houses where oils, attars, absolutes, and concretes are produced. The whole of the garden's design demands utmost care and sensitivity to season, so that at any moment during the year, many kinds of fragrant plants set their scents upon the air, at different hours of the day. Among them are the night-flowering Nicotianas, Moonflowers, Brugmansias, and Jasmines, thus to perpetually scent the night.

With perfumery, the old principle of concealing the stink of poor personal hygiene is, in theory, transcended to the art of subtle attraction, intrigue, and seduction. Yet this is redolent, in some of the most basic ways, to the replication of animal spoor, itself the very fetor of the 'uncivilized.' As with so many human arts then, it is less a matter of rejecting Nature in favor of the primacy of the human, but rather re-arranging it in forms that fascinate and please. Should we doubt this, we note that the vast number of plant sources in commercial perfumes are often balanced by concentrated odors from animals. The art of perfumes teaches us not only the immediate properties of aromatic construction, but how they blend with the natural musks of the body, how that collaborative odor transforms over time as it is volatilized, and how it acts upon others. Beyond the body itself, our esoteric knowledge of perfumes understands the occult properties of aroma, and what scents are pleasing to the spirits and gods, and those malodorous plant aromas best used for the expungements of infectious entities.

The Garden of Restoration

As with the byways of the Asklepion, medicinal plants dwell here in this healing land, where herbs themselves are the doctors, teaching mankind their powers, as did old Chiron. Essential plantings of Chamomile, Sanguisorbia, Yarrow, and Mandragora commingle in their beds with Cotton, Yerba Santa, and, their merciful mistress, Opium, in an endless variety of colors. For its soothing and mucilaginous virtues, all tribes of Mallows make their home near an impluvium for the cleansing of the body, and to draw water for sacred baths. In addition to the artfully designed palaces of diagnosis, treatment, and convalescence, the Garden of Restoration boasts a dreaming *tholos*, so that that patients may, in the night-revelations of sleep, seek the intervention of the god in their cures. From pre-natal care of mother and child, to birth, adulthood, and hospice care, each cycle of life, and the needs of the body, is attended by a magico-medical Arcanum in which plant powers play a direct role.

It is a matter of debate to what extent the modern elimination of the spiritual component of medicine has led industrialized humanity to its varied crises in health care. Indubitably, there still flourish societies where the spirit approach remains a vital, if not central, component of medicine. The pressures and persecutions that these groups have endured, on multiple fronts, is an abomination, but we may look to them as preserving the dignity of the most ancient bequests of healing, as well as the actual plant species that might otherwise have perished. Yet in some current models, the tide is turning; psychology, as a discipline, was a needful initial step toward restoration of the spirit. At the present time, the more progressive of medical organizations are allowing patients access to traditional spirit healers, alternative therapies, and chaplains ordained in the rites of minority faiths as part of a comprehensive medical approach to the individual. The Garden of Restoration may yet build itself anew, but owing to the spiritual complexity of disease, patient, and plant therapeutic, it will always have an occult component.

The Garden of Embattlement

A vast and tangled Blackthorn thicket encompasses the Garden of Mars, devoted to conflict, conquest, battle, triumph, and destruction of the enemy. The foundation of its landscape is a series of massive earthworks fashioned to resemble battle ramparts and fortifications, each successive embankment supporting herbs and trees used in warfare, both magical and mundane. At the center of the garden a temple to bloody Nergal looms above the thick-set hedges, carven of red onyx. Built in the appearance of a siege tower, it rises amid a gnarled thicket of Zilla, Acacia, Tribulus, and Crucifixion Thorn; only the hardiest specimens of humanity enter its pathways, returning bloodied and punctured. Several expanses of Tobacco fan out in ingenious plantings, making use of red, white, and yellow-flowered varieties to form the fantastic image of Huitzilopochtli, Aztec Lord of War. Stewarding the best cultivars, these treasured beds are grown from the seed of ancient plants, the smoke of whose leaves sent death to enemies at all points of the compass. Nettles, Poison Oak, Chile Peppers, and irritating plants of every kind ward the hidden by-ways where new and horrific forms of warfare are developed by adepts of this martial science. A grim processional avenue is lined with crosses of Thorn and ancient Dule trees, from which the corpses of the vanquished hang in various states of decomposition.

The magic of overcoming an enemy, sometimes referred to as 'combat magic,' is likely one of the earliest forms of sorcery, and botanical materials and powers have long been valued for this Art. This includes the use of plants as weaponry, for killing, maiming, and causing confusion, as well as cursing and coercion. The varied South American traditional formulations of *curare* poison and their use, for example, fall into this category, as do 'soldier's charms'—the plants carried into battle for safety, and to avert weapons. Attesting to the survival of magic into the modern era are a corpus of magical charms placed on firearms, either to ensure they would always hit their marks, or curses to deprive a gun of its accuracy, many of which include herbs. Just a few of the plants in this category include Aspen, Black Cohosh, Boneset, Fennel, Hyssop, Scarlet Pimpernel, Tormentil, and Woodruff, as well as the great and blessed Laurel, tree of triumph.

The Garden of Many Mantles

Looking down from above, the garden's multicolored geometry would appear to the sky-gods as being patterned on the spokes of a Spinning Wheel. Also called the Garden of the Distaff, within these fair grounds are taught the charms of Na'amah the Weaver of ancient days, and her skills of loom, as well as the precise composition of the threads of the Morai, who spun out the fabric of reality. The fine textiles proceeding thence, aside from the fabric of the everyday wardrobe, are used for the ritual robe, the Veils of the Temple, altar cloths and magical pouches, the soft linens of the connubial bed, the witching dolls and brightly-patterned prayer rugs, and the burial shroud. Laid out in orderly plots are fields of the finest Flax, Cotton, and Ramie, all hedged about by Mulberry, that noble tree which provides food for the silkworm colonies and the magnificent *Tapa* cloth of the Polynesians. The central enclosure, a place of initiation for those who dwell here, is the Grove of the Veil, where plants providing ready cloth with their fresh leaves and bark offer a pleasing shade amid strangely mirrored atone altars. This sanctuary is inspired by the deep green conclaves of Eden, that place of primordial mystery where fair Eve and Adam once girded themselves with leaves to conceal their glory from the peeping eyes of a pervert god. Aside from modesty and privacy, this contemplative aspect of the garden provides the teachings of concealing and revealing, for each is an act of power. Together with the plants providing the fibers themselves, the garden fosters herbs producing the tints of the thousand-colored rainbow, colors with which to dye our threads. Along a cobbled path bordered by all kinds of mallows, a riotous procession of nymphs parades the splendor of these divine habiliments, donning colorful finery ranging from gowns of simple leaves and flowers, to the finest chiffon and linen.

The plants which gave fiber to humankind, like those producing cosmetics, are infrequently afforded magical or occult consideration, but their history is as rich and spiritually potent as the doctor's balsam and the witch's poison. Like the secret herbal traditions of midwifery, it was a tradition of women practitioners. Beyond the considerations of the material objects and processes themselves, the art of spinning, weaving, dyeing, and fabrication bears a close relationship with ritual, for the mantling of the body, in various layers, is a cipher for the progressive strata of the cosmos itself, as is the making of cloth from thread and yarn. How one veils oneself, and unveils oneself in relation to the world, is an act of self-determination and personal aesthesis, but also a sophisticated understanding about the mediation between Self and Other. Also of great importance in this sphere of botanical magic are the enduring sorceries of flesh-binding, knot-tying, and the power of the red thread.

The Garden of Smoke and Fire

It is situated in a vast circular crater, the collapsed earthen remains of a dormant volcano. This locus affords the sound geometry of the magic circle to enclose the garden, evokes enduring resonance with the chthonian gods of heat and fire, and resembles in its essential form the shape of brazier, bowl, and tripod. Along the south-facing slopes of this garden is a great avenue of balsam-bearing trees: Guggul, Mastic, Storax, Terebinth, Myrrh, Olibanum, and Piñon, each providing a blessed resin for the harvest to come. Nearby stands a procession of stone *horrea*: here harvested plant materials are sorted and used to prepare incense, and postulants learn of the arts of flame and thurible. Growing in sunny plantings are groups of aromatic Ferulas, their noble crowns exuding pungent gum, as well as Cymbopogon, Cinnamon, Tobacco, Sage, and fine-leaved Desert Artemisia, all perfuming the air. On the distant crater rim, several plots are devoted to Ash-coppicing and the essential work of charcoal-making;

here the garden is heavily wooded and just beyond the kilns there is a stately marble chapel built, it is said, as a shrine to the *carbonari*. Ancient volcanic sulfur deposits are also mined at another site, the odorous golden dust used to enrich incense formulae for summoning infernal spirits. And in the greatest extent of the north-western area of the crater floor, there unfolds an expansive cremation ground, where fires burn by day and night.

The vast arena of ceremony and lore concerning itself with fire and smoke, as produced by plants, gives life the Garden of Suffumigations. In folk magic and sorcery, such smoke is often (but not always) of a pleasing aroma, and is used for healing, blessing, and expunging—or calling forth—spirits of all kinds. In religion, it is an offering to please the gods. In both instances it serves as a medium of manifesting visions, and carrying forth prayers to the otherworld. Such is properly the work of incense: its formulation, study, and use, as well as the proper care and harvesting of those plants which provide their bodies as its material basis. Yet this discipline also includes maddening fumes and intoxicating vapors, as well as such divinatory methods as capnomancy, divination by smoke. Among the oldest of ritual and magical practices, incense is an offering proceeding from the ancient traditions of the Burnt Sacrifice. In such rituals, one learns of the peculiars of the altar, correct woods for the fire, prayers and incantations, and the gestures of dispersing the smoke.

The Garden of the Scribe

The garden unfolds as a book, page by page, letter by letter, to constellate the great color palette of the plant world, and the paper with which it conjoins to incept works of art, literature, and learning. The statues which preside here are of the Muses, and of Great Athena, Thoth, Itzamn, and all holy gods of alphabets, writing, poetry and eloquence. No mere idols, each awakens within its respective shrine in the form of inspiration, thereby to catalyze the flow of words and ideas to the human mind. Venerable stands of Cork Oak and Birch, as well as Mulberry and Paper Figs serve for the making of paper, and nearby, along the stream, are to be found prized varieties of ancient Papyrus. Here too, amid this wet ferment, are uncountable Reed species, each bringing its virtues for the making of the stylus. In a vast tract is the daunting 'field of a thousand inks' where thrive Pokeberry, Marigold, Coprinus, and a multitude of other herbs whose juices render the colors of creation. Deep within the garden is a scriptorium, and a vast archive. Here, it is said, reside the first books ever written, the tales of the origin of the magical cipher, and the many formularia the garden has produced over millennia.

The legacy of writing emerged in antiquity from the mineral, in incised stones, stelae, and clay tablets. Yet, since the emergence of papyrus and ink six thousand years ago, the great ingenium of the plant world and its products has supplanted the scribe's reliance on hard materials, and by its physical properties alone allowed greater dimensions of aesthetic, portability, storage, and expression. Importantly, some of the earliest writings in the vegetal medium were in fact books of spirits and magical formulae, uniting the realm of plant-derived lexica with the art of magic itself. Some of these old books of magic record formulae for magical inks, indexing various plant ingredients for tint and lightfastness, as well as venerated fixatives like Myrrh.

The Garden of Prophecy

The whole of its grounds is arranged by cunning gardeners to promote states of individual revelation; thus the form of the recursive labyrinth—the maze which folds in upon itself, generating a second maze—patterns the landscape. Indeed, tracing an ancient circuit in this garden's midst is an old turf-maze, where all contemplation begins and ends, and whose walkers dwell perpetually in the garden's mesmeric efflux. Likewise within these grounds are numerous spaces of isolation and deep subterranean chambers, into which hermits descend as *mystes* to undertake rites of communion. Among the many plants of power growing here, we find Harmal and Divinorum and the holy vine of the *ayahuascero*, and in the damp places, Great Teonanácatl. In the dry and sunny places there are vast populations of Peyotl that thrive without fear of molestation. In this place, the garden follies are numerous, each sublimely wrought and devoted to a different god or spirit of revelation. One of these, rising on a small tumulus, is a temple of Apollo Loxias, its dome and pillars carved of the finest Heliotropium. Surrounding it are Laurels and sunny swaths of Henbane, the acrid smoke of which drifts from the interior of the shrine. Another modest shrine, in the form of a large, altar-like boulder, is devoted to none other than Saint Peter, keeper of the Keys of Heaven. Rising near-by are glaucous columns of *Echinopsis pachanoi*, his sacred plant, flanked by all the herbs sacred to the ceremonies of this visionary cactus. Throughout the garden, the varied arts of botanomancy find a welcome home here, from flower divina-tion to the throwing of wooden staves to the reading of coffee grounds.

Many of the plants associated with the Garden of Prophecy give rise to states ordinarily characterized as derangement, but it is often the case in prophecy that madness so called is an essential component of oracular reception. Where proph-ecy is concerned, we are wise to remember that an essential precondition for seeking knowledge of the future is our com-plete ignorance of it. Dwelling within the present, it is difficult enough to know the condition of *now*, especially considering simultaneity of multiple events and their trajectories of possi-bility. Thus to gain knowledge of the future, one must in effect be transposed to it, through atemporal means—hence the way of plant prophecy. From the perspective of the prophet, this involves communion with entities whose essences, as man-ifest in the mind, are completely unlike the human psyche, yet able, through the Eucharistic process, to effect dialogue, even if lacking language as usually reckoned. As witnessed by the observer external to the process, this collusion of the two perspectives produces a distortion in the temporal body, even as the prophet and the informing spirit attain clarity of visionary communion. This understanding, no matter how useful, does not begin to address the matter of returning to the present time and place with the omens received. As some have said, the gaining of omens is the easy part—their correct interpretation is the true labor. As within the garden, so with-out: the labyrinth of revelation is recursive.

The Garden of Alchemy

The Garden of Alchemy arranges itself by divine order, a mirror of Macrocosm: its concentric walkways are lined with Lady's Mantle and Blessed Melissa, the better to remind the Adept of the principles of magnification. Seven plots, arranged as the points on a star, are consecrated to the seven planets: in each of these, herbs with corresponding celestial powers flourish, concentrating those heavenly virtues for the transmutation by Art. In the fragrant plot of Sol are found Citrus trees, Marigold, and Sundew, and all those herbs producing philosophical warmth, nourishing to the body of mankind. Luna's plot is shaded by Willows, and interspersed with ponds and fountains, where Lilies, Lotus, Reed, and Iris flourish, together with globose-fruited plants and those pungent, aromatic roots who flourish in *terra paludosa*. That plot belonging to Mercury contains Cretan Dittany, Dill, and Foxglove; its earth is planted with the noble Hazel, which in days of old gave its power the Staff of Hermes. In the bed of Venus, we find a pleasant orchard of Apple, Pear, and Quince, as well as rosaceous berries and a diverse troop of fragrant Artemisias. The punishing acre of Mars is home to the Barberry and Mezereon, together with Nettle and rambling Eglantine, and those plants governing the blood, its rousing and spilling. In the orderly plot of Jupiter, where one may be rejuvenated, there thrive nut trees of various kinds, as well as Fig, Sage, and Mullein. And in the septenary plot dwell Mandrake, Hellebore, Squill, and other powerful herbs of restriction, together with the great Yew, Saturn's own tree. In the center of this stellate labyrinth, sealed within a vine-covered chapel, is the laboratory, overseen by the visage of Hermes thrice-potent, wherein the realization of the outer division of the garden is effected.

At its heart, the Royal Art of Alchemy is the spiritualization of the processes of Nature, and an attempt to understand each of them in both perpetual and temporal contexts. It understands the respective natures of fixity and volatility, and thus exalts change as a deific form, yet one with which mankind can collaborate, if it is willing to accept its inevitability and explicit dynamics. The alchemical adept values the virtues of experimentation and discovery as essential to the operation. It is a matter of fact that many who tend their furnaces in secret insist that without these features, it is not truly Art: for Dame Alchymia to attend, the seeker must be willing not only to discover, but to admit ignorance as a precondition of revelation. This, of course, is the difference between the empty vessel and that which contains our Proof. As regards the dominion of plants, the relevant branch of Alchemy is the vegetable work, magnified through its apotheosis as *spagyria*.

Conclusion

As presently explicated, the arrangement of both Pathways and Gardens reveals an esoteric model borne of personal experience and immersive observation of how plants emanate power in diverse spiritual milieus. Not all pathways will be appropriate for each seeker, and some may approach by multiple routes, as best suits the complexity of individual need. As study commences, each practitioner will in turn understand his or her own relation to the gardens and ways of approach; and will in turn discover their own forks in the path, the byways of private revelation.

The occult architectonic which presently conceives plant mysteries as dynamic structures reminds us that the essential functions of the garden are beauty, private contemplation, and the aesthetic arrangement of the exterior *domus*. Likewise history provides us with gardens manifesting from a philosophical and religious impetus—the great legacy imperial religions, and the enclosed grounds of the monasteries. Later philosophies of ordering, such as the French Garden style, witnessed the emergence of the garden as the dominance of the human mind over Nature, a triumphalist stance resonant with much of Western Occultism. Subsequent philosophies of the garden sought to reverse this, adopting the ancient ordering of Nature as the primary architectural datum, arguably culminating in Permaculture, which not only provides sufficient resources for the gardener, but also the wildlife in adjacent biomes. In the present age, with considerable disparity in land ownership and the destruction of the Commons, the garden is largely about stewardship and access to plant resources of one's own choosing. Each of these animating principles is applicable to the study of occult herbalism, and of course there are many more.

The Pathways and Gardens, however, are but two parts of a Trinity of powers comprising our Art. They are not the same as the knowledge of Occult Herbalism itself, which succeeds or fails based on the virtues of the third factor: the Seeker. And, just as the Seeker, each route and enclosure is unique: when embodied as one, they reveal the serpent-track of Samael through the Edenic Labyrinth, the Green Basilica of the Woodwose through which all *viriditas* courses. To those who seek, the primal impulse is the return to Paradise, and the dark substrates of its mystery.

Transmission of Esoteric Plant Knowledge in the Twenty-First Century

'ESOTERIC PLANT KNOWLEDGE' is a term I have constructed specifically to refer to metaphysical aspects of botanical knowledge that are preserved, utilized, and perpetuated within a group or tradition. This knowledge can include the magical, spiritual, and folkloric dimensions of plants as well as the specifics of their use, and specialized knowledge about their material aspects. Because the word 'esoteric' implies a hidden, marginal, or interior knowledge, or that beyond which is usually understood or accepted, I have also used the term *occult herbalism* to refer to this. The words 'esoteric,'

'occult,' and 'magical' all further imply concepts of conceal-
ment, mysticism, and spiritual power, the very particulars of
a mystery tradition.

At the present time, what this knowledge consists of and
how it is transmitted has yet to be examined in a meaningful
way, especially within the ambit of western esotericism; nor
are there useful discussions of the essence of its transmission.
In the present article I will broadly circumscribe the metapar-
adigm of esoteric plant knowledge, with special attention to
how and why this knowledge is transmitted between individuals and groups. In the process, I will touch upon historical
and transcultural models of transmission, with the intent of
developing an occult botanical pedagogy.

Background and
Development of Rubric

A dual involvement in the disciplines of botany and traditional folk magic informs my perspective on esoteric plant
knowledge. As I am an allopathic herbalist and a practitioner
of folk magic in 'occult' contexts, this sphere of inquiry is one
that I have formally been immersed in since the early 1990s,
and have helped to develop, through practice and teaching,
in addition to a number of original essays and books on the
subject.[1] My educational background is principally twofold,
focusing on practical agricultural systems and models of

---

1 *Ars Philtron* (Chelmsford: Xoanon, 2000); *Viridarium Umbris*
(Chelmsford: Xoanon, 2005); *Veneficium: Magic, Witchcraft, and
the Poison Path* (Three Hands Press, 2012); and *The Green Mysteries*
(Three Hands Press, 2017).

organic farming and horticulture, and in the other instance on ethnobotany, particularly that of medieval Europe.

In terms of my esoteric background, I have since 2001 been an initiate of the Sabbatic Craft Tradition, a group which is presently the steward of a number of historical magical practices, beliefs, teachings, and customs which coalesced into their present forms in the late 1800s in Britain, and are ritually passed from master to prentice using a direct initiation rite. Although syncretic in nature, the main components of these practices were derived from English and Welsh folk magic, and used an advanced symbolism of the Witches' Sabbat as the patterning for its rites, spells, and general cosmology. Incorporating both Christian and non-Christian ritual, it also includes healing charms, lore concerning the Fallen Angels, and other teachings derived from Continental, Romany, and North American sources. The traditional practices of the group bear a slight resemblance to the traditional Christian sorcery of the Pennsylvania Germans known as *braucherei*; in addition this tradition possesses a strong component of plant charms, lore, recipes, and magic. Although the tradition is insular and accepts members by invitation only, and its interior work is not discussed openly by its practitioners, several of them, including myself, have published works which convey a part of its essence.[2] In formal contexts I have also studied Traditional Chinese Medicine and spagyric alchemy.

Additionally, a number of groundbreaking works have served in the development of my ideas regarding occult herbalism, and should be acknowledged. The first of these is Mooney and Fowler's prophetic 1990 book *Shattering: Food, Politics, and the Loss of Genetic Diversity*. This book went large-

---

2 Andrew Chumbley, *Azoëtia: A Grimoire of the Sabbatic Craft.*

ly unnoticed among students of western esotericism at the time of its publication, but as I was deeply immersed in the botanical world at the time, its applications to magical plant knowledge, and the survival of traditional magic at a broader level, were obvious.[3] The work documents the history of human control over plant populations by development of such propagation techniques as seed collection, as well as the ongoing and well-funded effort of large chemical, bio-technological, and agricultural corporate interests to control the food supply, alter plant genetics, and disrupt traditional farming methods, eroding species diversity in the process. The research also focuses on the critical role played by plants in open pollination, and that of traditional farmers in an increasingly mechanized agricultural world. Although geared primarily for a botanical and agricultural audience, the work offers several lessons about transmission for the magical practitioner.

Also influential was the *Pharmako* series of books by Dale Pendell, which utilizes the rubrics of science, alchemy, magic, and poetry in a precise and artful manner to conjure the mysteries of psychoactive plants. This triad of works, in itself a fascinating study in the dynamics of the transmission of plant power, is exemplary in its phytognostic insight. Perhaps more than this, the example set by the author through his direct immersion in his subject matter hearkens to the spirit of discovery, an original animus of Science.

---

3 Cary Fowler and Pat Mooney, *Shattering: Food, Politics and the Loss of Genetic Diversity*.

Esoteric Plant Knowledge

Historical examples of esoteric plant knowledge are numerous and spread widely across time and space. Given the broadest possible scope of definition, one might trace esoteric plant knowledge to the appearance of the chloroplast, or even further back to its ancestors the cyanobacteria, whose appearance two and a half billion years ago heralded the dawn of photosynthesis. This biological alchemy transformed Primordial Earth, releasing untold amounts of oxygen, fundamentally altering the planet's atmosphere, oceans, and crust, but more importantly incepted the plant dominion. The revolutionary function of the chloroplast, its symbiosis, and its perpetuation over time can easily be used as a mnemonic for esoteric plant knowledge.

In terms of understanding human traditions, we must turn our attention to the most ancient known rudiments of this art. In written sources, for example, we may discern the magico-medical tradition of the Hittites in cuneiform tablets such as the Bogazköy Anti-witchcraft therapies. This assembly of incantations and prescriptions invokes the powers of specific plants for curing, especially for driving away demonic forces and malefic sorcery. A typical cure follows:

If a man is bewitched, you steep *suadu*-plant, *hasû*-plant, *nuhurtu*-plant, and salt in water. You leave it out overnight under the stars. In the morning you strain it. You have him drink it on an empty stomach, and he will recover.[4]

---

4 Tzvi Abusch and Daniel Schwemer, *Corpus of Mesopotamian Anti-Witchcraft Rituals*, 79.

Although the term 'occult herbalism' would not precisely apply to this spell in a strict historical context, it is easily grasped that the medical paradigm underlying it includes what in the modern era would be occult forces (witchcraft) as intrinsic to the disease pathology, in addition to the suggestion of astrology as a component of the medicinal preparation. Although scientific forms of medicine as currently practiced in industrialized countries would discount the two 'non-scientific' components of this cure in modern diagnosis, many traditional methods of healing endure which would not. Other historical examples of esoteric plant knowledge in written form include Egyptian Medical Papyri, Greek Magical Papyri, as well as various medicinal and magical treatises.

More closely related to modern forms of occult tradition and practice is the *grimoire* corpus of Europe, a magical literary tradition with its roots in Greek, Latin, Egyptian, Jewish, and Islamic magic. Usually understood to be books of conjurations of demons and spirits, the books may be compilations of magical formulae, such as the *Svartkonstböcker* of the Scandinavian countries, or more conceptually developed theoretical manuals such as *The Lesser Key of Solomon*. A feature of some of these volumes is that possession of the book itself is to be regarded as identical with the transmission of its power, a principle also present with *The Long Lost Friend*, the 1820 book of Christian sorcery collecting the spells of the Pennsylvania Germans. Although the grimoire tradition is largely focused on spirit-conjuration, many of the texts contain important information on herbal magic, such as the composition of magical incenses using toxic botanical ingredients.

In the instance of textual sources for magical plant traditions, it is essential for the researcher and practitioner to understand that reification into written form represents a single record of a single aspect of a tradition, not the tradition itself. Lost are the practical applications and informed contemporary personal interpretations of what was inscribed, as well as the more immediate circumstances of the writer and the written. This essential lacuna in knowledge would otherwise be filled with direct teaching and practice of the knowledge in question. A second caveat lies in the fact that, although the written word has long been perceived to lend legitimacy and power to the subjects it treats, it is not always a pristine vehicle for the transmission of esoteric plant knowledge.[5]

When considering the historical record, apart from textual recension we must also consider the archaeological record. Here it is instructive to study plants discovered *in situ*, where they were actually being used. A consummate example is the Copper Age European man known as the Iceman or Ötzi, whose remarkably mummified corpse was discovered in the Ötzal Alps in 1991, and was accompanied by a number of plants whose speculative functions ranged from medicine to kindling fires. The plants associated with this archaeological find comprise 121 individual botanical taxa,[6] a particular-

---

5 A modern example is the popular 1971 publication *The Anarchist Cookbook* (Lyle Stuart), which contained crude recipes for extraction of psychoactive plant alkaloids alongside formulae for high explosives and chemicals with false claims of psychoactive properties. Though not 'magical,' *The Anarchist Cookbook* may be broadly defined as 'occult' and with these qualifications, can rightfully be considered a modern grimoire.

6 Andreas G. Heiss and Klaus Oeggl, *The Plant macro-remains from the Iceman site (Tisenjoch, Italiian-Austrian border, eastern Alps): New results on the glacier mummy's environment*, 1.

ly rich trove of archaeobotanical information. Grouped as a whole, the preserved plants may be considered a primitive pharmacopeia or ethnoflora, but the traditions associated with them are lost and can only be guessed at, albeit with 'good' guesses in the presence of supportive data. Likewise, the intact tombs of the ancient Egyptian kings have yielded well-preserved botanical collections, casting light on ancient medicinal, ritual, and mortuary practices, as well as aspects of daily life such as food.[7]

Present-day esoteric plant knowledge survives in traditional cultures, with varying degrees of erosion. In some cases, knowledge of this loss is present alongside physical artifact and spiritual empowerment. One such example is the Little Water Medicine Society of the Seneca, whose powerful sacred medicine bundles remain, but the precise knowledge of how they were made, and the plants within them, has been lost. Nonetheless, the rites of the Society—in which the mysteries of the bundles are also preserved in the vehicle of song—continue, enabling various forms of transmission and spiritual continuity.[8] The syncretic Afro-Caribbean religions are another contemporary instance of traditional plant knowledge, where the usage of herbs in ritual practices is widespread and carries an immediate linkage to spiritual power.

In industrialized societies, in instances where links of traditional knowledge have been severed or eroded, plant knowledge is still handed down in secret, though lacking the cultural linkage to animist religion and other factors which

---

7 F. N. Hepper, *Pharaoh's Flowers: The Botanical Treasures of Tutankhamun.*

8 William N. Fenton, *The Little Water Medicine Society of the Senecas.*

often characterize more complete traditions of plant magic. An example in this regard in the United States would be the explosion of entheogenic research in the 1990s, and the perpetuation of such knowledge as cultivation, preparation, and use of power plants outside the scrutiny of the law. Though this comparatively new field of botanical inquiry cannot in most ways be compared to traditional teachings, it may be considered an outlier of the rubric because of the secret nature of the knowledge, the fact of its in-person transmission, and the centrality of plants as sources of power.

Historical Transmission of Plant Knowledge and Power

A known means of transmission of esoteric plant knowledge is the dispensation of herbal magic from spirits, gods, goddesses, fairies, mythic ancestors, or other preternatural beings directly to mankind in ancient times. This conveyance of power is present in myth, legend, and also in some origin-traditions of magic and sorcery. Positioning the origin of plant knowledge within the realm of the divine attests to its sacred nature, but also creates an immediate linkage between the magic practitioner and the gods, through an established chain of succession.

One such tradition is found in the *Book of Enoch*, a non-canonical Jewish text whose oldest sections are thought to date to circa 300 BCE. The book features a narrative of the Watchers, angels who descended to earth and took human wives and fathered the Nephilim, sometimes referred to as giants. Though this descent is characterized as a 'fall' there is also a willful element present among the angels, such as their intentional design to marry human wives, bound in compact with a sworn oath. After consummating their bond with human females, the Watchers taught various arts, skills, and species of magic to humanity. Semjâzâ, the leader of the fallen host, taught 'enchantments, and root-cuttings,' presumably a form of plant magic. Various versions of *Enoch* relate how this introduction of angelic arts 'befouled the earth' and otherwise introduced iniquity into the world.

Although assuming the form of scripture or myth, this telling contains interesting components relevant to our examination. In the Watchers narrative, not only was the transmission of magic a forbidden act, but the nature of the knowledge itself was forbidden as well. This implies a secret or heretical tradition, concealed because of its transgressive nature, similar in many ways to what passed between the Woman and the Serpent in the third chapter of the Book of Genesis.

Shadows of the story of the Watchers, or fallen angels, can be found in the witch persecutions of late medieval and early modern Europe, specifically in the descriptions of the Witches' grand ritual, the Sabbat. This phantasmagoric revel, typically painted in orgiastic or abominable hues by its legal detractors, typically involved inversions of Christian ritual, which later became known as the Black Mass. Scholarship long held that the phenomenon was an example of mass hysteria, or clerical confabulation, but neglected to examine the axis of actual folklore, folk belief, and magical practices. In classic demonological descriptions of the Sabbat, the Devil often presided at the rite, and, often engaged in horrifying or obscene acts with the witches. Curiously, in some accounts he also taught the use of making poisons, ointments, or powders.

The micro-historian Carlo Ginzburg was one of the first scholars to dissect the lesser known components of the Witches' Sabbat, due to anomalies found in court records of the Friuli. His study led to the discovery of a deep stratum of non-Christian folk-belief in Europe which resonated with, and was sometimes assimilated to the features of the Sabbat.[9] Among these dwelt the *benandanti*, a group of quasi-shamanic magical practitioners who went forth by night to battle in spirit for the fertility of the fields. The female patron of the Sabbat in its Romanian form, was known as Doamna Zînelor, and was known as the mistress of the fairies. At the ritual, which was sometimes attended in dream, she was known to teach the magical secrets of herbs; Ginzburg importantly identifies her as a goddess of the dead.[10] Other lady-patrons of Sabbat such as Irodeasa, Arada, and also Doamna Zînelor, are still associated with the ritual dance and ecstatic frenzy of the Romanian *căluşari*, who are considered both magicians and healers.

Set against the same backdrop witch-persecution was the Sicilian Fairy-Cult of the *Donas de Fuera*. This group consisted predominantly of women but also included men; they were charismatic magical healers by day that undertook nocturnal travels and engaged in ceremonial reverie which contained many classical elements of the Sabbat. Court records reflect that those who admitted to having traveled to the Sabbat noted the presence of fairies in the guise of theriomorphs, ap-

---

9 Carlo Ginzburg, *Ecstasies: Deciphering the Witches' Sabbath*. Moshe Idel, a scholar of Esoteric Judaism, has also recently argued convincingly for a Jewish contribution to elements of the medieval Witches' Sabbat in his book *Saturn's Jews: On the Witches' Sabbat and Sabbateanism*.

10 Ginzburg, Ibid., 71–77.

pearing as beautiful humans, but having the feet of animals such as cats paws or horse's hooves. Typical activities were paying fealty to the King and Queen of the fairies, singing, feasting, and sexual pleasure. When contrasted with the more standard and abhorrent and malefic forms of the Sabbat in Inquisitional literature, the Sabbats of the *Donas de Fuera* are more centered around empowerment of the celebrants and the attainment of various forms of ecstasy. In particular the transmission of herbal cures and practices was in some cases a part of the ritual itself, and was connected in the waking world with the Donas' healing practice.[11] Whether the Sabbat took on an evil form or a more beneficent one, it is clear that it stands as a mythic strand of transmission of plant lore, with emphasis on a supernatural origin, a hidden tradition, and the presence of aspirants receiving the tutelage.

The fairy-doctors of medieval Wales, whose traditional cures incorporated elements of allopathic herbalism and occult practices, perpetuated a healing tradition transmitted through family lines. The *Physicians of Myddfai*, a manual of magical cures recorded in the eighteenth century but by tradition said to date from the 1200s, features, among other receipts, healing charms using an apple inscribed with a nail:

> *For all sorts of ague, write in three apples on three separate days.*
> *In the first apple O Nagla Pater*
> *In the second apple O Nagla Filius*
> *In the Third Apple O Nagla Spiritus Sanctus.*
> *And on the third day he will recover.*[12]

---

11 Gustav Henningsen, "The Ladies From Outside: An Archaic Pattern of the Witches Sabbath." 191–215.

12 *The Physicians of Myddfai.*

As noted, the magic of the Welsh *Meddygon Myddfai* was hereditary, passed down in a familial bloodline, as well as a secondary route through text. Additionally, the family ascribed a spirit-origin to the plant knowledge, which ultimately originated with a fairy woman in a local lake. This route of transmission of plant magic from the fairies is also historically attested with Huw Lloyd, the fifteenth-century Welsh wizard whose power was attributed in part to his possession of magical books given to him by a fairy woman who lived in Bridge Lake.

As a further instance concerning the transmission of knowledge from water, the syncretic *curanderismo* traditions of Peru, in which plant powers and knowledge play a central role, offer an instructional perspective. The practical part of the tradition is largely comprised of magical curing, involving prayer, song, sprinkling, offerings for spirit-intercession, invocation of saints and pre-Columbian spirits, and the use of various magical plant preparations, including the vision-producing cactus *Echinopsis pachanoi*. A critical part of the tradition involves regular pilgrimages to sacred highland lagoons, such as those in Huancabamba, Mishahuanga, and Páramo Blanco, whose waters serve as a repository of the curers' power. These pilgrimages to the lagoons and their immediate environs also serve as a means of magical transmission, as new charms are taught and learned there, specifically relating to, and emanating from, features of the location. Herbs critical for the work are also collected.[13]

The power of a *genius loci* to transmit magical knowledge, in this case lakes and lagoons, reminds one of the Meddygon Myddfai and the lake-dwelling fairy who taught magical

---

13 Donald Joralemon and Douglas Sharon, *Sorcery and Shamanism: Curanderos and Clients in Northern Peru.*

healing. Aside from the geophysical similarity, the Peruvian lagoons also bear some similarities to the enchantments of fairy-lore. In one instance a curandero reported the appearance of an unfolding vision in one of the lagoons:

> At that point it was seen that the lagoon was transformed, that is, one was looking at the water, and one didn't see water anymore. Instead, one could see houses—because the lagoon is immense, it's big—one could see houses and people inside.[14]

Another *curandero*, upon witnessing this, suddenly rushed into the water, and had to be dragged out, lest he be trapped and remain within the water, where a great serpent dwelt. Although this account took place in 1972, it is reminiscent of the medieval British account of King Herla, who violated the taboos of the dwarf-king of the underworld, and as a result was doomed to eternally wander; an alternate telling has Herla and his band plunge into the River Wye, never to be seen again.[15]

Within the complex of folk traditions that in unity comprise the modern Sabbatic Witchcraft Tradition, there exists a sacrificial ritual which is sometimes referred to as 'The Rite of Making Green.' This rite was first written about in a redacted form, by Andrew D. Chumbley, and it was one of many such rites I undertook in order to become that tradition's Verdelet, the steward of its traditional plant-lore.[16] The aspirant jour-

---

14 Ibid., 91.

15 Walter Map, *De Nugis Curialium*, c. 1140–1200.

16 Andrew Chumbley, 'The Secret Nature of Ritual.'

neys alone to a remote and forlorn woodland and stands beneath the trees in silence. One's clothing is cast aside in sacrifice, and one stands naked before the arboreal host, despite wind, weather, insects or other travail. Sap, earth, and leaves are gathered up from the site and rubbed into the skin from head to toe, effectively mantling the body. Thus guised, the aspirant invokes the spirit of the trees and runs without heed rock, thorn, or branch until a state of gnosis ensues. Although this rite is more complex than I have described here, and must be assimilated within the interior context of the Sabbatic Tradition, I can attest to its power. The rite stands not only as an example of esoteric plant knowledge, but also indeed of the process of transmission, not only for its heuristic route of knowledge but also for the linkage between the aspirant and the *genius loci*.

Impediments to Transmission

Our examination of the contours of transmission of esoteric plant knowledge would be incomplete without a brief consideration of its obstacles. In doing so I will not examine legal, religious, and industrial forces, but instead focus upon the act of the transmission and its spiritual heart: the teacher, the student, and the knowledge itself.

The concept of *false transmission* indicates the transmission of alleged spiritual teachings of a traditional nature without the authority to do so. Within this category of dubious persons we locate the 'plastic medicine man,' a pejorative term applied those who ape or appropriate of Native American culture, claiming to be shamans.[17] Characteristic of this type of individual is the lack of actual connection to the culture being appropriated, as well as use of the teachings for ego or money. Although the plastic medicine man is most readily apparent (and most commonly repudiated) in the New Age movement, the phenomenon is also present in occult and esoteric organizations.

A kindred concept is the transmission of falsehood, in other words, the teaching of corrupt and inaccurate information. Aside from running the risk of having one's public reputation destroyed, this presents real-world dangers to students, particularly if the teachings being conveyed involve toxic plants, or magical techniques that place the body under physical duress.

Exploitation of resources for personal monetary gain can also affect the integrity of transmission. A classic example where this occurs is an herb walk. Throughout the United States, Canada, England, and in Europe, nature walks have in some areas become increasingly focused in the past decades on wild plants. Whether identifying local medicinal plants, or a practicum on wild edibles, these excursions can be useful for orienting the seeker in the Book of Nature, and reading from it in a practical way. But quality of education and expertise can vary, and such events are often a platform for promotion of a personal product line, or are exorbitantly priced. As just one of many instances relating to experiences with herb walks, in 1992, I formally severed contact with one of my plant teachers after I watched him harvest an entire population of Licorice Fern (*Polypodium glycyrrhiza*) on a so-

---

17 David Chidester, *Authentic Fakes: Religion and American Popular Culture*, 173.

called 'wildcrafting' instructional walk. Not only was the act harmful to the plant population, it transmitted the idea that doing so was acceptable.

Additionally we must acknowledge that certain individuals, due to irresponsible behavior, mental instability, or other contraindications, are wholly unsuited for transmission. Accidents or misuse of plant power can have severe consequences, as well as legal implications that affect all practitioners, not to mention perpetuation of ignorant stereotypes. In discussing these types of individuals, Dale Pendell invokes the horrific visual emblem of the Ten of Swords tarot card, and counsels us to 'never share poisons with the unlucky.'[18]

Proposed Future Models

Speaking from my own experience, and having directly observed the situation of others over time, I have found that there is no substitute for direct one-on-one teaching. Indeed, the ancient method of oral instruction is particularly suited for working with plants. In addition to allowing for the subtleties of direct instruction and the ability to answer questions directly, it allows a unique bond between mentor and prentice which is characterized by unique attitudes and approaches. Thus it is not purely knowledge and facts that are transmitted but also nuance, perspective, practicalities, and the accumulated chain of initiatic power—a school of thought.

Such interpersonal tutelary relationships are exceedingly rare, and their cost is great in time and treasure. For persons of industrialized societies, without deep ties to traditional

---

18 Dale Pendell, *Pharmako/Poeia*, 81–82.

cultures, the question of honest and meaningful access to esoteric plant knowledge will naturally arise. Aside from formal
education, and independent scholarly study, there are several
pathways of individual transmission.

The first form is direct learning from Nature itself, through
observation. Known as *heuresis*, meaning discovery, this differs from strict empiricism, and one must be careful to examine one's motives, and in the process of observation avoid the
imposition of personal fantasy. Put more simply, this mode
route of transmission and reception might simply be called
'paying attention.' Nonetheless, if undertaken sincerely, it
provides information which is immediately useful to the individual.

A basic form of understanding of the plant world that one
can gain in this fashion is the creation of a personal phenological plant calendar. Essentially, this is a record kept from
direct observation of plants, season after season, year after
year, for a given locality. Recorded within the calendar are
such data as a complete flora of the given area, population
densities, when certain species flower and set fruit, weather
and how it affects the plants, diseases, and plant succession
over the years. One might also include the observed relation
between the studied plants and animals: what serves as food
for birds, what plant resources are impacted by deer and ro-
dent populations, and so forth. Expanding the investigation
yet further, one can also note phenotypic differences from
one population to the next. For example, many wild huck-

leberry populations may inhabit a certain habitat, but some shrub groupings many have larger berries, or more abundant ones, or none at all. These respective data points, if faithfully recorded, will provide the practitioner with abundant carto-graphic data about his or her nearest wilderness, its inhabitants, and their actual nature, but also their relationship to the land and to time. More importantly, because the survey data is tied to locality, the knowledge gained will be directly relevant to the practitioner. In such cases the transmission is effective when a private understanding of the plants of the locale as a whole is gained, and the relationship between practitioner and plants cemented through personal understanding, practice, and stewardship. As a vocational extension of this, the British model of the gamewarden or *verderer* is exemplary.

Other basic forms of 'informal' transmission in postmodern industrial society include the passing of private family recipes through time, working in community gardens and gaining knowledge of plants from fellow participants, and other engagements within one's own community. Beyond those directly known, one's local agricultural extension or Native Plant Society is a worthy place to start; although such groups may not be wholly 'occult' in nature, one often finds in their ranks one or two individuals with a mystical perspective on the botanical.

Hearkening to the necessity of the *grimoire* or magical manual, consideration should also be given to the private formulary. In essence, this is a book of one's own private esoteric work, being recipes, charms, spells, and teachings that one uses or for other reasons is deemed worthy of preserving.

Here it will be observed that the rationales for a compiling a private book of formulae, intended solely for private use or to pass on to a chosen individual, will differ significantly than those for the book which is composed for public viewing. In essence, money and notoriety are removed from the equation, and the focus remains on the fitness of the transmission, and also of those who will receive and use it. All practitioners who keep such records have an obligation to plan for their ultimate fates, whether transmission to human hands or to a bonfire. Examples of the latter are a well-honored tradition in the occult arts and sciences.

In consideration of the foregoing routes, I would also like to point out the benefits of intentional community to esoteric plant knowledge and its transmission, with a personal example. In 1994, I and several other individuals with kindred interests formed the plant research group Peninsula in Santa Cruz, California. Many factors and currents drove the structure of the group, but for myself one such influence was the 'Temporary Autonomous Zone' of Hakim Bey, a theory of forming highly functional collectives of limited duration outside usual and formal hierarchical structures.[19] The purpose of the group was to explore psychoactive plants in a manner that encouraged direct discovery, but also the combined routes of knowledge of all group members. The group included among its ranks a botanist, a surgeon, a computer programmer, an acupuncturist, a musician, a rave DJ, a psychiatric counselor, a poet, and a florist, among many others. The basic format of the work consisted of a single individual ingesting a chosen psychoactive plant substance, and in a subsequent meeting reporting back concerning the experience. In this manner, the diversity of perspective and vocation in the group was utilized as a strength and served to generate an atmosphere

---

19 Hakim Bey (Peter Lamborn Wilson), *T.A.Z.: The Temporary Autonomous Zone.* The book was highly influential for political activists in the 1990s and remains a relevant resource today.

of multidisciplinary analysis, as a supportive background to individual discovery.

Intentional community, as an organizing principle, is an appropriate model for the transmission of esoteric plant knowledge because of its ancient legacy of organizing a group of individuals around concerns of plants, whether it be of the hunter-gatherer paradigm or some form of agriculture. Furthermore, I would also note that unity in the form of a group, with common philosophical goals, also resonates with certain aspects of plant communities in the wild.

Conclusion

Though taking many forms historically, the metaphysics of the act of transmitting magical knowledge about plants can be expressed in three simple variables:

- The essence of the teacher;
- The essence of the student;
- The essence of the transmission.

If these variables are all of high integrity, the act of transmission is likely to be successful. However if one or more are lacking, transmission is incomplete, corrupt, or nonexistent. Ultimately, the proof lies within the carrier of the current, who must not only exemplify its teachings in a theoretical fashion, but also in a practical one. Such is the difference between the static 'dead letter' and the 'living book.'

Teachers—those with the authority to transmit—in addition to standing as exemplars and embodiments of that which is taught, must also exemplify the tutorial ideal; if one cannot admit fallibility, egalitarian discourse is impossible. The student too must be worthy of that which is to be taught, knowing the virtues of silence, reflection, and patience, but also understanding one's own unique powers and gifts, even if they lie steps away from complete realization.

Above all, once transmission of knowledge has occurred, there should be no doubt of this, as demonstrated by the one who has received the transmission. Such proof lies far beyond the exercise of memorization by rote; it must be lived and demonstrated. The proof lies in the Work.

Occult Herbalism:
Ethos, Praxis, and
Spirit-Congress

OCCULT HERBALISM is the sorcerous art of calling forth power from plant spirits, their devotional worship, and mystical understanding. An active spirit-bond with the Greenwood bound by votive praxis, it is positioned within almost every religion and tradition of magic, whether presently vivified or forming the part of the magical substrate of antiquity. One living exemplar is the Curandero traditions of Peru, which regard plants as spirit forms, as much as any saint or god. Another is spagyric alchemy and iatrochemistry of early modern Europe, which sought an herb's exalted quintessence through the formula *Solve et Coagula, et habebis mag-isterium*—'dissolve and bind, and you shall master.' In many spirit-streams of the African diaspora, such as Ifa, Santeria, and Vodun, the herbal component is not isolated, but wholly integrated into religious practice. Closer in essence to my own background as an herbal practitioner is the *wortcunning* of Anglo-Saxon leechcraft, distinct artifacts of which are found in modern Essex cunning-craft and indeed elsewhere in rural Britain. Similarly, in North America, where traditions of folk-magical herbalism began arriving from Europe and England in the sixteenth century, Occult herbalism proceeds as a many-trunked tree, hybridized with the ancient stock of Native American ceremonial medicine and the potent plants sorceries of Mother Africa. Hoodoo, with its compelling stock of powders, oils, washes, and botanical talismans, represents one such amalgam, but there are many others, including the yarb-doctors of the Ozarks, the Mexican *brujos* of the southern United States, the granny-doctors, and the Appalachian pellars. Importantly, all traditions of occult herbalism bear three components which, like a close-knit effusion of branches and vines, weave a thicket of power accessible to the magical practitioner.

Ethos

The first of these strands is ethos, presently defined as 'the character, sentiment, or disposition of an individual or a group considered as a natural endowment,' and also: 'the spirit which actuates manners and customs.' It may rightly be said that every artifact, book, philosophy, group, and individual radiates an ethos to those who may rightly discern it. Whether consciously known or acknowledged, ethos remains an actuating force or magical engine driving every creation to act present and cognize as it does. There is also an ethos present in every form of magic. Consider the following prayer, taken from an ancient charm for harvesting plants:

Your size is equal to the zenith of Helios,
Your roots come from the depths,
But your powers, they are the heart of Hermes,
Your fibers are the bones of Mnevis,
And your flowers are the eye of Horus,
Your seed the seed of Pan.
I wash you in resin as I also wash the gods
Even as I do this for my own health.
I am Hermes, I acquire you
With Good Fortune and Good Daimon
Both at a propitious hour
And on a propitious day effective for all things.

This is an abbreviation of a plant-harvesting rite appearing in the Greek Magical Papyri, a large corpus of spells and emanating from Hellenized Egypt, and ultimately reflecting components of classical Greek and Egyptian religion, but also Gnostic and Judaeo-Christian magical concerns. This invocation is spoken directly by the harvester, to the plant he is about to cut; it follows a ritual of self-purification and burnt offering of the incense kyphi. In the rite there is a delicacy of approach present acknowledging the power and centrality of the plant genius; respect is also evident by calling forth by name the praiseworthy attributes of the plant. In attendance also is the realization that the operation is potentially danger-ous and must be undertaken with proper apotropaic proce-dures in place, such as cleansing oneself with resin.

Also found within the charm is acknowledgment that the magician has entered the precinct of power, marking a shift from ordinary to a state of hallowed or sacred Otherness. The sorcerer additionally calls forth the spirit-lineage of this plant, associating it with a retinue of gods, all of which are benevolent, life-giving, and illuminating. One particular god summoned, Hermes, was the patron of the Magical Art itself and he is here co-identified with the practitioner, an assumption of deity possibly employed as an amuletic ward. Taking this spell's ethos in totality, we may summarize its stance of power as applied animism. Despite its mention of the mortal gods, their virtues are not anthropomorphized. Rather, the incantation approaches as sentient and animate that which moderns often regard as inanimate. The charm thus regards the plant to be harvested as a fellow, a contem-porary, and an emissary of power. In this there is an implied hierarchy of Seeker and One-who-gives-Power. We may also witness similar ethe present in a 'Charm to be spoken over a Divining Rod':

I, X, beseech you Noble Stick,
In the name of the Father,
The Son,
And the Holy Spirit,
That you truthfully show me,
Point out, and guide me to the place
Where there is gold, silver, or money,
Which has been made by human hands
And is hidden in the earth.
Be as true as the Virgin Mary
Was a pure virgin,
Before and after the birth of Christ.
So certain and true as Christ is the Son of God,
And so certain and true
As God's son died for man's sins
And is all mankind's savior.
I beseech you, stick,
With your might and power
As Jesus conquered the Devil
And all evil spirits and made them most humble.
I beseech you, stick,
That you alone point out old treasures which
   are hidden by man, and not any other minerals or metals.
Continue looking when you are commanded.
With God's holy gospel,
His holy word,
With the angels, the martyrs,
   the Devil's downfall and confinement,
And with a loud trumpet sound that on the Day of Judgment
   shall give life to all Nature's dead and living creatures
   that have been on the earth.

This spell, an abbreviation of a slightly longer form, is separated from the Graeco-Egyptian spell by about 1,700 years; it comes from a *Svartkonstbok* or 'book of black arts' found in Norway dating to between 1790 and 1820 and recently discovered in an old farmhouse. Published nominally as *The Black Books of Elverum* (Galde Press, Inc., 2006), it is one of many such books that have survived into the present day and are held in archives and private collections. In ethos, it mirrors certain formulae utilized in traditional British witchcraft, such as a syncretism of Christian elements with an older substratum of non-Christian rhabdomancy.

Though separated in time by almost two millennia, and by magico-religious context, many of the characteristics of the Egyptian and Norwegian charms are identical. First, by addressing the plant as a living, sentient entity, much in the same manner as addressing a god. Second, the quality of the devotion is similar in both charms, praising the plant and invoking its noble qualities. The words are specifically chosen to magnify the magical power of the wood. This implies a hierarchy of *divinity*, namely the plant and the sorcerer. Third, both magical operations beseech the plant for power, which implies the greatest humility, as well as the implicit acceptance that the plant's answer might be in the negative. Another feature both charms share is oath-binding in the names of the practitioner's most holy gods. In the case of the Norwegian spell, despite the obviously devoted Christian practice of the writer, this charm invokes an essentially polytheistic retinue: the Saints, the Martyrs, the Holy Angels, God the Father, and Jesus the Son.

One other ethos present in most well-developed traditions of plant magic bears mention: a proximal sorcerous relationship to one's chosen magical pharmacopoeia as a route of witching power. This specifically concerns addressing the plant directly in the templum of its own habitation, a markedly different approach than ordering one's plant material pre-harvested through the post. A proximal relationship implies sowing and tending the plant in one's own garden; or, where it can be done responsibly, going into the wild to gather it in accordance with taboo. It should also be stressed that even among the oldest folk magical traditions, one finds imported plant materials in use, having accreting their own 'lore of the exotic.' The ancient ingression of Frankincense into English folk magic, both high and low, was almost certainly due to its associations the Magi and the magical use by the Church to purify. Local practitioners likely adopted it and contextualized its use, not only in resonance with the Church's sorcerous ethos, but with the deeper substrate of historical 'folk religiosity' with which they were already familiar.

One cannot, therefore, apply an exclusive 'prejudice of locality' to the entirety of plants in any given magical tradition, though the majority of magical plant traditions with historical continuity I am personally familiar with use mostly local plants, harvested by the practitioner, and steward unique and highly precise knowledge about those herbs. A proximal relationship to plants thus yields a magical pharmacopoeia which is fresh, ready to hand and linked in space, time, and lore to its source thus keeping the power of the *genius loci* contextualized.

Additional reasons for the ethos of proximal relationship include the magical act of pilgrimage—the process of going to the plant, petitioning for its power and harvesting. If one intentionally sets out upon the path by ritual declaration of intent, one 'lays the field' for the procession of omen and sign to come. The endpoint of pilgrimage is not simply obtaining

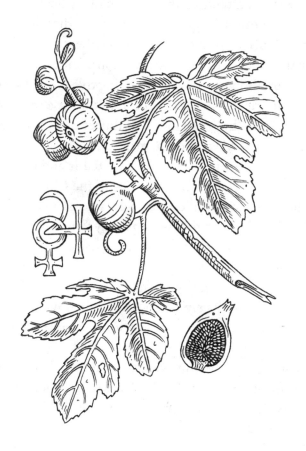

what one set out for, but the mystical understanding indwelling every step going forth and returning.

Praxis

As a second cartographic dimension of Occult Herbalism, praxis is the spiritual action of the practitioner, the inception of magical activity. Here, the archaic form of the word is used, the better to distinguish it from the drudgery and repetition accreted in the modern era to the word 'practice.' Praxis may thus be understood as the regular devotion of magical word, deed, and numen to the spirits. In part, praxis naturally arises out of ethos. To grasp this in a meaningful way, in order for praxis to proceed, the practitioner must first define his or her ethos, a deed easier said than done: where aspiration conforms to actuation, one may fairly define one's ethos. If a sincere answer, without pretense, can be divined, part of the basis of praxis can be glimpsed with new eyes. For many, discovering one's ethos can be a frightening exercise, for upon flaying the untold layers of psyche and spirit, it is often discovered that much of a practitioner's ethos was previously unknown and in fact contradicts one's aspirations on the path. Thus a rigorous, honest exercise in in spiritual deconstruction is indicated, as well as the work of self-slaying in sacrifice to the magical art.

As a broad example of the praxes of Occult Herbalism, it is instructive to examine the perennial phenomenon of plant-harvesting taboos. Anyone who has seriously studied plant magic in traditional contexts will encounter these Laws of the Greenwood, which deliberately address questions such as time of harvesting, ritual purity of the practitioner, appropriate offerings to the spirits, and correct magical uses of the plant material gathered. All tessellate to form the law of plant taboo.

Casting our eyes upon British folk magic and its established streams of occult herbalism, a persistent stratum of praxis emerges that can be defined as a magical traditional based largely (but not exclusively) on fairies, their magic, and appeasement. A great wealth of plant harvesting and planting taboos concern appropriate demeanor between humans and the Good Folk, including ways of gaining their favor, respecting their dominions, and avoiding their wrath and subsequent malice. This notion of treading lightly about the Folk of Elphame is still pervasive even among Christians and secular folk; academic analysis might identify this as a vein of superstition, though in actuality such is quite often suffused with a deep reverence approaching religion. Practices such as planting certain trees such as Rowan (*Sorbus acuparia*) at boundaries, offerings of food set out at night, and fear of cutting specified woods such as Elder (*Sambucus nigra*) still command great respect. The plant taboos of fairy magic thus circumscribe a boundary governing acceptable human interaction with plants and their warden spirits, and exist as much for the protection of mankind as for the respect of the Folk of Elphame. Giving due consideration to customary taboo is thus a respectful and appropriate place for the practitioner to begin defining a magical praxis with any given plant.

Even if one regards plant taboos with complete skepticism, or as superstition, there are still good reasons to honor them. In the first instance, such adherence honors the plant as well as the human culture that relates to it and gave rise to the taboo. Aside from occult decorum, there may be good ecological reasons for respecting a taboo of harvesting, such as limiting harvest to avoid depletion of slow-growing species. Beyond this, to dismiss plant taboo outright goes beyond being inconsiderate, it approaches hubris, for taboo in itself is a field of instruction, wherein one can gain entirely new fields of knowledge and perspective by stepping outside the selfish constraints of personal impulse and desire.

For those whose who aspire to the art of occult herbalism, there is a praxis which, if undertaken in sincerity and humility can incept a shift in sorcerous perception necessary for attenuating congress with plant spirits. Its essence is to seek and obtain a branch of totemic wood, originating from one's own resonant arboreal familiar, to serve as Ally in all magical work. As such, alignment of one's own *numen* with the arboreal famulus must be achieved, this may only happen after rigorous self-examination of one's own magical ethos. This Great Work accomplished, the practitioner then identifies the points of personal resonance with the sanctified tree. Blackthorn (*Prunus spinosa*) and Whitethorn (*Crataegus monogyna*), though thorny and sharing identical habitat, each possess a very different animating power. Practitioners aligned with the Yew (*Taxus baccata*), especially in its wild state, will know its unique atmosphere, from which the ethos of the tree spirit may be determined singular and clear. Those who revere the Pine will affirm that its spirit-resonance is complete unlike the Spruce, though both are evergreen conifers.

Undertake pilgrimage: set out from the place of beginning, go prayerfully unto the tree, abide, and return. The specifics of the journey must remain unknown, because ultimately it is each practitioner's pilgrimage to make: the spirits will attend, or not, in accord with their favor or displeasure. At journey's end, the practitioner may have little more than he began with, or may in fact possess the golden distillate of the Circle thrice-wandered. Upon reaching the tree of pilgrimage, acknowledge your patron, by actuation of magical word and deed, as Ally: a deified being unto whom sacrifice is freely given. Within its immediate proximity discern the continuum of power that lies betwixt healing and harming—where a plant with strong healing affinities is sought, know also its power in opposition, and how it may be used to harm. Where a wood for cursing is required, know and acknowledge how that same wood may heal. Such may be done in meditation, prayer, invocation, and by other means of Art as ingenium guides. If all is well augured, respectfully cut a small branch-length to be retained as the prime arboreal liaison of instruction—the Magistellus of Wood. Thence, let it pass within one's private devotional shrine, to become part of daily contemplative practices.

Through the alchemy of time, praxis, and innate understanding the raw branch may assume the form of wand, key, stave, idol, mask, or olisbos, in accordance with the magical aims of the practitioner. The wood will have thus moved from activity, through contextualized empowerment, to activity anew: the hidden deeds of Art. In this manner the occult herbalist receives the impress of the Tree's power in omina, having set foot in the Circle of Green. This reflects motions from the potentials of ethos, through the activity of praxis, into the dynamism of *spirit-congress* which lies at the heart of occult herbalism.

Spirit-Congress

Concerning the spirit-congress, the third part of our formula of occult herbalism, we best define it as direct interact between practitioner and plant power by means of sorcery, of which there are many forms. In folklore, such often includes a dispensation of power from the dominion of Elphame, sometimes sealed by the marriage of the magician to a fairy-wife of husband. The *dyn hysbys*, a particular type of Welsh cunning-man known in times past, is exemplary in this respect. Though at times also known as 'white witches' or 'faith healers' many exemplified the dual-hand ethos of curing and cursing, seen in acts of magical warfare known as *unbewitching*. Some were reputed to have gained their power from fairies, for example the fifteenth-century dyn hysbys Huw Llwyd, who lived at the foot of the Coxcomb climb below Mt. Snowdon. His power lie not only in the fact that he was a learned doctor, but that he possessed books of magical learning. Furthermore, his two great grimoires were said to have been gained directly from a fairy woman who resided in the nearby Bridge Lake. It is said that before he died these books were thrown back into the lake, where they were reclaimed by the hands of the fairy woman who dwelt there. Similar legends are found throughout Wales. Perhaps better known is the tale of the Lady of Llyn-y-fan-vâch near the vale of Myddfai, Carmarthenshire, who spawned a generation of powerful doctors of half-human half-fairy blood, and a unique Welsh tradition of highly advanced magical healing.

Here it is important to consider that in the historical context of vernacular religiosity, to identify oneself as a good Christian would not have excluded the acceptance of fairies, angels, astrology, incantations, charms, and the direct intercession of spirits. Indeed, if one looks at the 1743 Welsh manuscript *Meddygon Myddfai*, one finds medicine, magic, astrology, Christian faith, and fairy lore as conspecifics.

Ethos, Praxis, and Spirit Congress are three elements recurring in traditions of occult herbalism; they may be thought of as a threefold formula of sorcerous autocatalysis. Each component represents a zone of power, a field which, rightly tended, will yield up the grain of power and knowledge. Though these principles are applied herein to the Art of occult herbalism, they apply equally to any sphere of power, and are especially apropos Marriage: the veneration of, and union with, the Beloved.

Ethos governs how the Beloved-of-Art is cognized, summoned, encountered, and reified. It is always present as a first principle, whether conscious or not, and whether acknowledged or not. It cannot be changed quickly on a personal level, and arises more from the spiritual path one has walked, fair or foul, than from the path one craves but has not walked. It is often the case that magical aspiration and ethos are unknowingly in conflict, and sabotage the work. Praxis, which arises out of Ethos, governs the active cultivation and nurturance of the bond with the Beloved, but also its strengthening or weakening. In addition to rigor of exaction and breadth of inspiration, praxis must be sincere, devoted, and in accordance with the guidance of the heart. Spirit-Congress is the goal of the work, in the real (rather than imagined) intercourse of practitioner and plant power. It is the place where Desire meets Fate, and where one may in truth call the Beloved 'Patron, Familiar, and God.' Should it come to pass, let the good herbalist beware: marriage unto spirit is no less demanding and challenging than a marriage in the flesh!

The Green Intercessor:
Tutelary Spirits and the
Transmission of Plant-Magic

And all the others together with them took unto themselves
wives, and each chose for himself one, and they began to go in
and defile themselves with them, and they taught them charms
and enchantments, and the cutting of roots, and made themselves
acquainted with plants.

I Enoch 7:1–2

IN THE INTRODUCTION to his 1970 book *Mastering*
Witchcraft, British author Paul Huson ascribes the origin of
witchcraft to 'that mysterious dark angelic fire which first
breathed life into the clay of this world,' citing the so-called

fallen angels as the intercessors of this power.[1] While not the first to identify these luminaries as the source of witch-magic, the book was unique in its stance of sympathy toward the practitioner of witchcraft (and the exiled angels), as well as its receipt of wide distribution. This inclusion of fallen angel lore may be considered a counterpoint to contemporary books defining witchcraft as 'Wicca' or as opposed to other 'traditional' currents of witch-lore, and to contemporary Wiccans' uneasy relationship with the Judaic components of their religion.

Huson quotes from the third-century *Book of Enoch*, which gives a descriptive account of how a group of rebellious angels called Watchers were exiled from Heaven, married human women, and taught the arts of magic. This story, also found in Genesis 6:2 and various extra-biblical sources, was widely read in ancient times and was also adopted by Christianity, though the Coptic Church is the main congregation which presently regards it as canonical. The Enochic lore of angelic descent was spread by early authors such as Origen, Irenæus, and Tertullian, who ascribed the origins of evil to these beings.

Augustine, who with great relish expanded the fallen angel narrative to broaden Satanic influence upon it, was ironically a Manichæan before his conversion—a Gnostic religion whose canonical *Book of Giants* has recently been shown to be derived from, or share a common origin with, the Watchers sections of *Enoch*.[2]

---

1 Paul Huson, *Mastering Witchcraft* pp. 11–20. Charles Godfrey Leland's controversial *Aradia: Gospel of the Witches* (London: David Nutt, 1899), considered a foundation of the modern witchcraft revival, must also be considered for its inclusion of Lucifer.

2 John C. Reeves, *Jewish Lore in Manichaean Cosmogony: Studies in the Book of Giants Traditions.*

In addition to the angels' fall, the *Book of Enoch* records about twenty specific names of the most important angelic commanders. Several of the angels are named as patrons of specific types of magic taught to humankind, including geomancy and astrology. One magical art so taught was the knowledge of plants and the 'dividing of roots', attributed in some sources to Shemhazai, the leader of the reprobate host.[3]

These matters are of personal interest to me because as a practitioner of folk magic I share a background in occult herbalism and traditional witchcraft, as well as being an initiate of the mysteries and faith of Christianity. The former acknowledges an active spiritual dimension to plants and herb-magic; the latter has transmitted to me both occult herb-lore and magical teachings concerning the Watchers. My perspective is thus that of an occult practitioner interested in the historical and spiritual cartography of certain aspects of these arts, specifically the transmission of magical plant knowledge from angelic or spirituous entities. In the absence of a satisfying theological or anthropological term which encapsulates the dynamic of magical knowledge of plant origin, I have in previous writings called it *phytognosis*, or gnosis arising from plants.[4] However, before considering the daimon of the plant itself as the intercessor of direct occult power, I would like to examine the role of angels and other spirit-beings.

---

3 In variant texts the leader is also named Azazel, goat-headed patron of metallurgy, cosmetic adornment, and sorcery in general.

4 Daniel A. Schulke, *Ars Philtron* p. 9.

In Jewish lore, the Watchers were also called the Sons of God, or *Bene Älohim*, and were similar in some ways to the Muslim *djinn*, created of smokeless fire and inferior to other angels, who are said to be created of light. The fourteenth-century *Chronicles of Jehremeel* of Eleazar ben Asher ha-Levi speaks of the Watcher Shemhazai (Manichaean *Shemhazad*) as the revealer of the ineffable name of God. This he spoke to the human woman Estirah; upon uttering it, she ascended, and at that moment became fixed in the Pleiades.[5] This juxtaposition of the giving of plant-knowledge and the revelation of the Most Holy Name bears similar transgressive contours to Samael's revelation of the powers of the Tree of Knowledge of Good and Evil to Eve, the First Woman. Both instances share the common feature of plant powers (forbidden fruit, or in the case of Shemhazai, the knowledge of plants and the division of roots) given by a wayward angel to a human woman, as well as the a direct and personal transgress against God.

As bearers of magical knowledge, some angels are regarded in Jewish lore as dream-giving intermediaries. Sometimes called 'The Dispenser of Dreams,' the *Memuneh* is the guardian angel of man 'who molds his sleeping thoughts to apprise him of the will of God.'[6] In some cases such dreams involved the direct transmission of knowledge from the plant realm. Consider the following extract from the Dead Sea Scrolls:

> I, Abram, had a dream the night of my entry into the land of Egypt. In my dream I saw a cedar tree and a date palm growing from a single root. The people came intending to cut down and uproot the cedar, thereby to leave the date

---

5 *The Chronicles of Jehremeel.*

6 Joshua Tractenberg, *Jewish Magic and Superstition*, p. 235.

palm by itself. The date palm, however, objected, and said, 'do not cut the cedar down, for the two of us grow from but a single root.' So the cedar was spared because of the date palm, and was not cut down.[7]

The notion of secret knowledge delivered by angels to humanity in Jewish lore also occurs in non-transgressive contexts. *The Book of the Angel Rezial*, said to have been delivered by to Adam by the Angel Rezial, contained a number of treatises on angelology, kabbalah, gematria, and astrology. In this cycle of lore, the authorship of the magical knowledge is said to be God himself.[8]

A modern exemplar with curious parallels to the lore of the Watchers is the *Zar* spirit-possession cult of North Africa, whose practitioners traffic with djinn-like beings. *Zar* (pl. *zayran*) refers to the possessing spirit, though the faithful refer to the spirits as *asyad* or 'masters.' The *Zar* ceremony contains ritual features of ecstatic dance, music, and singing, ritual suffumigations, and spirit-offerings including animal sacrifice, resembling in many ways the *bembé* or *mange loa* of Vodun, as well as the primal ritual patterning of the Witches' Sabbat. Though it appears the *zayran* are not regarded by all practitioners as *djinn*, they are endowed with several of their attributes, such incorporeality and the ability to possess the human body. In other demonological-angelogical constructs *zayran* constitute a third class of *djinn* (after the benevolent 'white *djinn*' and malevolent 'black *djinn*') which are red in color, characterized as capricious, ambivalent, and

---

7 *The Dead Sea Scrolls: A New Translation*, ed. Michael Wise, Martin Abegg, and Edward Cook, p. 79.

8 *Sepher Rezial Hemelach.*

pleasure-seeking. Recalling the legend of the Watchers, most *zayran* are considered to be male, and their spirit-mediums of 'brides' are female; indeed the *Zar* cult is dominated by women. Additionally the *zayran* fulfill a tutelary function within the rites of possession, revealing specific songs. Gerda Singers, who studies the cult in Cairo, notes that in present-day Egypt, understanding of the nature of demons includes beneficient and malevolent beings; good djinn are referred to as 'earthly angels.'[9]

Christianity, with its complex and sometimes contradictory traditions of demonology and angelology, degraded not only the heathen gods to the status of demon, but all fallen angels as well. In the syncretic streams of medieval and early modern European magic, these were to resurrect anew as the Goetic Spirits, the operative powers which in some cases animated the Solomonic systems of sorcery. Not surprisingly, the *Goetia* ascribed the powers of revealing plant knowledge to no less than five tutelary demons; the spirits Bathin, Morax, Foras, Stolas, and Bifrons specifically teach the virtues of herbs. A text linking the old Levant traditions of magic with the later European necromancer is *The Testament of Solomon*, dating from the third century CE. Here, Solomon compels the demon Asmodeus to reveal the power which thwarts him; he dictates a formula for a magical suffumigation using the liver and galls of a fish, and a branch of Storax.[10]

---

9 Gerda Sengers, *Women and Demons: Cult Healing in Islamic Egypt*, p. 103. Trance-possession cults invoking the *djinn* have been documented by Vincent Crapanzano in 'Saints, Jnun, and Dreams: An Essay in Moroccan Ethnopsychology,' *Psychiatry* 38 (1975).

10 *The Testament of Solomon*, 6:4.

Apart from the demonological and angelogical sources of Near Eastern origin, there are a number of tantalizing antecedents in European folk-magic for spirits revealing magical knowledge of herbs in ritual contexts. Some of these may partake of Jewish Watcher lore, but other cases clearly represent a spirit-cartography outside of that context. Carlo Ginzburg notes that in certain witchcraft trials in Ragusa in the latter half of the seventeenth century, defendants referred to themselves as Villenize and professed learning their curing and magical herb-arts from the Vile, who were arboreal and vegetal spirits.[11] The night-traveling *bona gens* or 'good people' were condemned by the Milanese Inquisition in the late fourteenth century for consorting with the spirit-divinity Madonna Hortiente, who taught her subjects 'The Art' and 'the efficacy of herbs.'[12] Wolfgang Behringer notes that traditions of folk-healing in Greece, Bulgaria, and Dalmatia recognize illnesses of supernatural origin, which can only be successfully treated by healers who have received their powers from supernatural beings such as nereids.[13] Concerning the spirit-cults of Eastern Europe, Éva Pócs cites an example of dream-derived herbal knowledge of Hungarian fairy heal-

---

11 Carlo Ginzburg, *The Night Battles: Witchcraft and Agrarian Cults in the Sixteenth & Seventeenth Centuries*, p. 142.

12 Wolfgang Behringer, *Shaman of Oberstdorf*, pp. 53–54.

13 Ibid., 88.

ers.[14] In Russian lore, wormwood (*Artemisia* spp.) is associated with Serpents, especially their queen, who bestows the 'power of speech and the uses of plants.'[15] Russia is the source of a legend explaining the origin of wood-sprites as remnants of legions of fallen angels.[16]

Gustav Henningsen has documented in Sicily an important night-traveling cult which represents in some ways an inversion of the diabolism typical of the European Witches' Sabbat. Resembling in many ways the oneiric spirit-cults investigated by Ginzburg, Behringer, and others, the 'fairy sabbaths' as described by its attendees bear some of the classic attributes of the witch-ceremony such as feasting, music, dancing, and venery. Also present were celebrants riding goats, a feature of the Sabbat similar to the riding of the sacrificial ram of the *Zar* cult. Yet these ritual components, as well as the *Donna di fuora* (the ethereal spirits in attendance) are cast in a beneficent and ecstatic light, contrary to usual Inquisition documentation. Notably absent are the cannibalism, obscenity, sexual pain, and ritualized Christian inversion common in many ecclesiastical accounts of the Sabbat.[17]

As evidenced by the previously-cited Russian lore, fairies and nature spirits have been traditionally explained as the progeny of the Fallen Angels. This is in accord with rabbinical tradition, which explains the spirits known as *lutins* and

---

14 Éva Pócs, *Between the Living and the Dead*, p. 153.

15 Charles M. Skinner, *Myths and Legends of Flowers, Trees, Fruits, and Plants*, p. 299.

16 W. F. Ryan, *The Bathhouse at Midnight*, p. 37.

17 Gustav Henningsen, 'The Ladies From Outside: An Archaic Pattern of the Witches' Sabbath,' pp. 203–4.

faes as the offspring of female demons and Adam, after his expulsion from Eden.[18] Perhaps more relevant to traditions of witchcraft in England, the *Carmina Gaedelica* records the lamp-lit song of the Hebridean fairies' dance:

> Not of the seed of Adam are we,
> Nor is Abraham our father,
> But of the seed of the Proud Angel,
> Driven forth from Heaven.

It bears mention that in much of British fairy-lore, the race of Elphame are not the pleasant and whimsical creatures of Victorian paintings but malevolent beings capable of bringing mischief, disease, and ill luck. Much of the rural magic involving them consists of propitiation to avoid human offense and protect farm and family. Some have been ascribed the quality of bestowing herbal knowledge to mortals, such as the Lady of Llyn-y-fan-Vach of Myddfai, from whom is derived an entire hereditary tradition of Welsh folk healing. Also present in this tradition, however, is the Lady's power to withdraw from congress with humans if not approached correctly.[19] Stories of the Fallen angels were also in popular currency in England from the Middle Ages on, bearing particular Old Anglo-Saxon features, such as the use of the word *scufan* ('shove') to describe the angels' expulsion from the celestial realms, or *wlite* ('brightness, beauty') to describe their radiance.[20]

---

18 Tractenberg, *Jewish Magic and Superstition*, p. 29.

19 Ann Ross, *Folklore of Wales*.

20 Catherine Brown Tkacz, 'Heaven and Fallen Angels in Old English' in *The Devil, Heresy, and Witchcraft in the Middle Ages*.

It would seem from cursory examination that the Watchers and fairies are part of an array of tutelary spirits providing magical knowledge to chosen humans, in some cases magical knowledge of plants. The difference in how this knowledge was recorded varies greatly, but this would be of little concern to the recipients of the intercessory power. In Enochic lore, the dispensation of such knowledge is ultimately regarded by the chroniclers as a transgression against God and a polluting influence upon humanity. In the case of the Sicilian fairy-sabbaths, the regents of the fairies give their faithful 'remedies for curing the sick,' and 'did not wish them to do evil things, but to heal [people].'[21]

Emma Wilby has made a thorough study of the classes of spirit-intercessors in early modern British witchcraft, particularly the witch's familiar, drawing parallels in the magical audience of such beings with shamanism.[22] Of particular interest to our study are the 'fairy familiars' of the cunning folk. The mid-seventeenth century exemplar of Anne Jeffries of Cornwall reveals that the tutelary relationship between spirit and magical practitioner involved not only medical diagnosis of illness but also prescribing cures. Records of the time indicate that Jeffries' renown was such that people came from as far away as London to be cured by her. [23]

The sexual component of angelic magical transmission is also of note, for like the Watchers account, records of the magical patterning of the Witches' Sabbat reveal sexual activity alongside the transmission of plant lore. Among the sab-

---

21 Gustav Henningsen, 'The Ladies from Outside,' 196.

22 Emma Wilby, *Cunning Folk and Familiar Spirits*.

23 Wilby, Ibid., 68.

batic rites of the *Donna di fuora*, the spirit-attendees engage in
festive 'love-making,' two features of which are the great pleasure and the frequency of the act. Henningsen contrasts this
depiction with the usual Inquisition narrative of depraved or-
gies and sexual congress with demons.[24] Though the genders
are reversed, this union of starry wisdom with the green earth
evokes the perpetual connubial embrace of the Egyptian sky
goddess Nut with the terrestrial god Geb, whose recumbent
body is often depicted green. The presence of sexual activity in
the context of transmission of plant knowledge may be a rem-
nant of sexual ritual activity within agrarian deity worship,
or may function, perhaps, as a cipher of the process of spirit
possession itself. The attainment of gnostic states via sexual
trance is also part of the teachings of the Sabbatic Tradition,
the branch of early modern magical practice into which I am
initiated. Indeed in some modern witchcraft recensions—as
well as other non witchcraft forms of traditional sorcery—the
transmission of power from Initiator to initiand occurs exclusively during a sexual rite, or can only be passed from female
to male or vice-versa.[25] Ida Craddock, the nineteenth-century
American occultist whose astral sexual relationship with the
angel Soph is detailed in her treatise *Heavenly Bridegrooms*,
made use of these astral formulae and *dynamis*. Aleister Crowley, in reviewing her work, affirmed the initiatic nature of the
knowledge she had attained, saying *Bridegrooms* was 'one of
the most remarkable human documents ever produced' and
was 'absolutely sane in every line.'[26]

---

24 Henningsen, Ibid., 197.

25 Vance Randolph, *Ozark Superstitions* pp. 266–7.

26 Ida Craddock, *Heavenly Bridegrooms*, p. 280.

Moving from an externalized source of wisdom-intercession, we may consider the plants themselves as angelic emissaries. Dale Pendell, who has written extensively about tutelary plant spirits, employs the term 'allies' as did Castaneda, to describe a plant's tutelary mask.[27] In my work I have often utilized *genii* in accord with the Latin components of some forms of witchcraft.

When considering the corpus of Watchers legends, we are obliged to seek the trees of Old Eden as tutelary divinities and reservoirs of spirit-knowledge. Trees, in particular those bearing nuts,[28] were considered in Jewish lore to be gathering places for demons; this recalls the notorious Walnut of Beneventum about which the *Striga* assembled and danced by night. *Sepher Hasidim* warns that trees bearing drippings resembling candle-wax, presumably exuded resin, are gathering places of the *liliot*, night-demons associated with Lilith, also an important figure in witchcraft. On a number of occasions, both in England and in America, I have been taken by teachers of widely-differing spiritual traditions to certain trees which I have since come to refer to as 'showing trees.' Their purpose, not initially disclosed to me, was to reveal certain aspects of a person when they stood beneath their branches. As such, the trees acted as a nexus of power serving to 'unclothe' the person or lower his psychic guards. With at least two examples, I have witnessed this phenomenon repeated times at the same tree, even when this was not the practitioner's purpose. Such individual arbors, usually of solitary grotesque form or considerable age, have included hawthorn, elder, and oaks.

---

27 Dale Pendell, *Pharmako/Poeia*.

28 Tractenberg, *Jewish Magic and Superstition*, p. 34.

In the context of my own magical practice, it is traditionally taught that the magical power of a tree or herb arises from its unique 'virtue' (*virtus*). These virtues comprise a distinct sphere of the plant's power and are conterminous with, but differ from, its spirit. Though this virtue is not specifically transmitted to the witch-practitioner by a spirit exterior to the plant (*daimon*, angel, etc.), the practitioner may 'borrow' or gain access to it by various forms of entreaty, ritual harvesting, and charm. Additionally, in the context of English folk magic, use of the archaism 'virtue' to describe plant powers is suggestive of something more. 'Virtue' has been occasionally used in traditional witchcraft circles to describe the quintessential witch's power, especially in the West Country. The term was so used by the traditional witch Robert Cochrane, who noted its transient and motile qualities, as well as its specialized witchcraft meaning beyond common usage.[29] The passage of virtue from one witch to another mirrors not only the egress of *virtus* from herb to practitioner, but also from tutelary spirit to sorcerer.

Plant spirits as the sources of direct transmission magical knowledge is a feature of the shamanic world, and rich traditions of plant-allies endure in the spiritual teachings of indigenous Americans. A South American corollary with the spirit flight or spirit-possession cults are the Peruvian specialists of plant magic known as the vegetalistas. Their ritual imbibition of the *ayahuasca* decoction for healing, divination, prophecy, spirit-travel, and necromantic communion is accompanied by spirit-emissaries known as 'little doctors.' These spirits convey specialized magical knowledge, especially in relation to the magical use of plants. Like the *zayran* of the *Zar* rites,

29 Robert Cochrane, 'A Witches Esbat.'

the little doctors also teach unique songs by which the spirits may be called and their power commanded.[30]

'Dream helpers,' which assist the shamans of many North American tribes, bear many similarities to the witch's familiar, and many of them arrive through the direct intercession of plants. Often the helper is an animal soul, such as Coyote, Frog, or Owl; but it can also be an elemental such as water. Like the witches' familiars, these intercessors provide specialized power to the shaman, such as the poisoning power provided by Rattlesnake. The Chumash people, as well as other California tribes, have long sought the tutelary spirits by drinking an infusion or decoction of root of Datura (*Datura* spp.), a potent visionary plant known in European witch-lore as thorn-apple. As with the little doctors, dream-helpers of the Datura-vision could convey unique magical songs, knowledge of plants, and specialized powers.[31]

In my own personal practice, I have identified many modalities of spirit-tutelage in working with plants, wherein the linkage between practitioner and spirit-intermediary occurs through specified 'roads' including dream, trance-possession, waking vision, votive offering, and consumption

---

30 Luis Eduardo Luna, 'The Concept of Plants as Teachers among four Mestizo Shamans of Iquitos, Northeastern Perú,' pp. 135–156.

31 Jan Timbrook, *Chumash Ethnobotany.* See also Pendell, D. *Pharmako/Gnosis.*

of plant corpora as an empowered eucharist. Each bears a unique signature of the plant in question, as well as a distinct manner of tutelage; one exemplar I will discuss here in brief. Though these signature are consistent with both old lore and modern reports, the experience of them within the circle of enchantment is unique to each practitioner.

A fortunate straying on my magical path many years ago led to instruction in the visionary use of Belladonna (*Atropa belladonna*), a plant I had long studied and grown, but, out of both fear and respect, never sacramentally imbibed. Belladonna is a plant of considerable toxicity and a member of the nightshade family, a group of plants with ancient ties to witchcraft in both the Old World and the New. My initial experience under the effects of the plant involved a compression of the physical sensorium. This sense of restriction is well grounded in biochemistry: hyoscyamine, a tropane alkaloid which Belladonna possesses in considerable concentration, is utilized in medicine for drying up excess fluid, and is responsible for the side effect of a parched mouth. In these initial stages of being overshadowed by the plant, I was unaware of any presence I would call 'spirituous' though a number of curious characteristics were noted, chiefly my inability to stay out of trance, the very definition of a narcotic. A sense of increasing restriction and numbness overcame me, and compression of my consciousness narrowed to a singular, finite point. The room was suddenly awash with shadows and I felt the weight of an ancient spirit-column above me, a great tangled mass of souls. The suddenness of its appearance and its monolithic nature further reduced my perception, leaving my immediate reckoning in a floating mist of anaesthetized terror. Before this black edifice I shuddered, beholding the moment of my own death, and the soul's eclipse. A sense of panic and terror, however, only arose when, in fear of expiration, I raised somatic defenses against the genius. However, when I embraced her awful power, and the inevitability of my imminent demise, the mask of the goddess shifted from wrathful to tutelary. It was as if the genius was insistent I behold her as she truly is, within the ambit of her own dominion, rather than what I had read and presumed.

For the next several hours my consciousness and corporeality were pierced by the flight of umbral shades swarming about a central 'column of atavism,' attended by a clamor of whispers, shrieks, and groans. Though the voices spoke in what sounded like an internally consistent language, it was incomprehensible to me but for its intuited qualities. Subsequent contemplations after this virginal congress seemed to indicate that the central black column was the shade of the Beautiful Lady herself, and the souls which formed her swarming mantles were those of practitioners, who in carnal form had used Belladonna magically. In subsequent years, the appearance of the Lady has shifted forms, like the glamour of the witch, but her manner of communication has not, marked as it is by great horror and beauty.

The spirit-intercessors of the herb sorcerer, lurking as they do in the green vales of every magical tradition, will no doubt continue to reveal themselves to their chosen, whether they be practitioners of the magical art or respectful academics. Within the purview of magical practice, it is hoped that it is not simply the source of gnosis that is emphasized, but the trinity of practitioner, point of emanation, and spiritual congress between them. The result of this magical trine is the ecstatic state known as the Witches' Sabbat, the co-mingling of souls and flesh wherein the gates of heaven and hell are thrown wide. Yet this is but one perspective, that of the practitioner of traditional witchcraft. Given its applicability to spirit-given plant knowledge, perhaps a better term for this state is Hildegard of Bingen's *Viriditas*, the divine vitality of the green world. At once applicable to the Ancient Ones of Light, first smitten by Woman, and the ethereal princes and queens of the Dominion of Faerie, it evokes both the power and mystical potentials of the hidden granary of Eden.

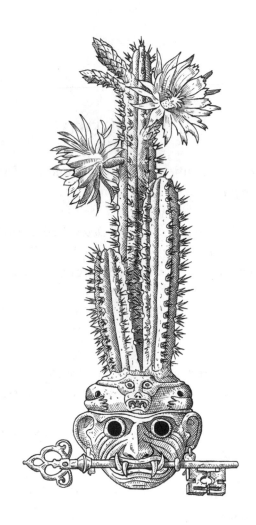

Bibliography

Abusch, Tzvi and Daniel Schwemer. *Corpus of Mesopotamian Anti-Witchcraft Rituals*. Brill: Leiden, 2011.

Behringer, Wolfgang. *Shaman of Oberstdorf: Chonrad Stoecklin and the Phantoms of the Night*, trans. H.C. Erik Midelfort. Charlottesville: University Press of Virginia, 1998.

Bey, Hakim. *T.A.Z.: The Temporary Autonomous Zone*. New York: Autonomedia, 1991.

Chidester, David. *Authentic Fakes: Religion and American Popular Culture*. Berkeley: University of California Press, 2005.

Chumbley, Andrew. *Azoëtia: A Grimoire of the Sabbatic Craft*. Chelmsford: Xoanon, 1992.

— 'The Secret Nature of Ritual,' *Chaos International* 18, London: 1995.

Cochrane, Robert. 'A Witches Esbat,' *New Dimensions*, 2:10, 1964/65.

Craddock, Ida. *Heavenly Bridegrooms: An Unintentional Contribution to the Erotogenic Interpretation of Religion*. New York, privately printed, 1918.

Crapanzano, Vincent. 'Saints, Jnun, and Dreams: An Essay in Moroccan Ethnopsychology'. *Psychiatry* 38, 1975.

Fenton, William N. *The Little Water Medicine Society of the Senecas*. Norman: University of Oklahoma Press, 2002.

Fowler, Cary and Pat Mooney. *Shattering: Food, Politics, and the Loss of Genetic Diversity*. Tucson: University of Arizona Press, 1990.

Gaster, M., trans. *The Chronicles of Jehremeel*. New York, Ktav Publishing, 1971.

Ginzburg, Carlo. *The Night Battles: Witchcraft and Agrarian Cults in the Sixteenth & Seventeenth Centuries*, trans. John and Anne Tedeschi. Baltimore, Johns Hopkins University Press, 1992.

Heiss, Andreas G. and Klaus Oeggl. *The Plant macro-remains from the Iceman Site (Tisenjoch, Italian-Austrian border, eastern Alps): New results on the glacier mummy's environment*. Berlin: Springer-Verlag, 1991.

Henningsen, Gustav. 'The Ladies From Outside: An Archaic Pattern of the Witches' Sabbath,' in *Early Modern Witchcraft: Centres and Peripheries*, ed. B. Ankarloo and G. Henningsen. Oxford: Oxford University Press, 2001.

Hepper, F. N. *Pharaoh's Flowers: The Botanical Treasures of Tutankhamun*. London: H. M. S. O., 1990.

Huson, Paul. *Mastering Witchcraft*. New York: G. P. Putnam's Sons, 1970.

Idel, Moshe. *Saturn's Jews: On the Witches' Sabbat and Sabbateanism*. London: Continuum, 2011.

Joralemon, Donald and Douglas Sharon, *Sorcery and Shamanism: Curanderos and Clients in Northern Peru*. Salt Lake City: University of Utah Press, 1993.

Luna, Luis Eduardo. 'The Concept of Plants as Teachers among four Mestizo Shamans of Iquitos, Northeastern Perú,' *The Journal of Ethnopharmacology* 11, 1984.

Map, Walter. *De Nugis Curialium*, c. 1140–1200.

Pendell, Dale. *Pharmako/Poeia*. San Francisco: Mercury House, 1996.

Pócs, Éva. *Between the Living and the Dead*. Budapest: Central European University Press, 1999.

Randolph, Vance. *Ozark Superstitions*. New York: Columbia University Press, 1947.

Reeves, John C. *Jewish Lore in Manichaean Cosmogony: Studies in the Book of Giants Traditions*. Cincinnati: Hebrew Union College, 1992.

Ross, Ann. *Folklore of Wales.* Stroud: Tempus, 2001.

Ryan, W. F. *The Bathhouse at Midnight: Magic in Russia.* University Park: Pennsylvania State University Press, 1999.

Rustad, Mary, ed. *The Black Books of Elverum.* Galde Press, Inc., 2006.

Savedow, Steve, ed. and trans. *Sepher Rezial Hemelach.* York Beach: Samuel Weiser, 2000.

Schulke, Daniel A. *Viridarium Umbris: The Pleasure-Garden of Shadow.* San Francisco: Xoanon Limited, 2005.

Sengers, Gerda. *Women and Demons: Cult Healing in Islamic Egypt.* Leiden: Brill, 2003.

Skinner, Charles M. *Myths and Legends of Flowers, Trees, Fruits, and Plants.* Philadelphia: J. B. Lippincott, 1911.

Tkacz, Catherine Brown. 'Heaven and Fallen Angels in Old English' in *The Devil, Heresy, and Witchcraft in the Middle Ages,* ed. Alberto Ferreiro. Boston: Brill, 1998.

Timbrook, Jan. *Chumash Ethnobotany.* Berkeley: Santa Barbara Museum of Natural History, 2007.

Tractenberg, Joshua. *Jewish Magic and Superstition.* Philadelphia: University of Pennsylvania Press, 2004 [1939].

Wilby, Emma. *Cunning Folk and Familiar Spirits.* Brighton: Sussex Academic Press, 2005.

Wise, Michael, Martin Abegg, and Edward Cook, eds. *The Dead Sea Scrolls: A New Translation.* San Francisco: Harper-Collins, 1996.

Index

Jupiter 73

katharsis 16–8

Mandrake 62, 73
Mars 65, 73
Meddygon Myddfai 88–9,
  111, 121
Memuneh 116
Mercury 73
Monkshood 21

Na'amah 66
Nephilim 85

Pendell, Dale 79, 93, 125, 127
Pennsylvania Dutch 78, 81

pharmahuasca 41
phenology 32, 34
purification 17–8, 101

Rose 13, 24, 27, 52, 58, 62
Rowan 108

Sabbat of the Witches 46,
  78, 86–8, 89, 90–91, 117,
  120, 123
Samael 75, 116
Saturn 39, 73, 87
Seneca peoples 83
Sensorium 54–7, 128
sexual calibration 49
Semjaza 86
Solomonic Magic 14, 118
Storax 57, 62, 118

suicide 57
Svartkonstböcker 81
syncretism 40, 104

taboo, plant 40, 90, 105,
  107–9
trance 9, 61, 118, 123, 127–8

Venus 59, 63, 73
vincula 23
Virgin, Pathway of 16–18

Watchers 85–6, 114–8, 122,
  125
Whitethorn 109
wildcrafting 93
wilderness 9, 11, 29–30, 60,
  95
witches' familiars 109, 112,
  122, 127
witchcraft 14, 46, 59, 62, 77,
  80–1, 90, 113, 121–3
wysoccan 47

Yew 57, 60, 73, 109

zar cult 117–20

Thirteen Pathways of Occult Herbalism
was published in May 2017 by Three
Hands Press. This first printing is
comprised of 3228 copies in total. Of
this are 2000 trade paper editions
printed offset and sewn with color
covers, 1200 standard hardcover
copies bound in green cloth with color
dust jacket; and 28 hand-numbered
copies quarter-bound in brown goat-
skin and autumn marbled paper.

SCRIBÆ QUO MYSTERIUM FAMULATUR

CPSIA information can be obtained
at www.ICGtesting.com
Printed in the USA
JSHW021649110321
12445JS00002B/2